Colour reproduction in the printing industry

Anthony Mortimer

© Pira International, August 1991

ISBN 0 902799 76 2 (hardback)

ISBN 1 85802 012 3 (paperback)

Published by

Pira International

Printed in England by

Antony Rowe Ltd, Chippenham, Wiltshire

For sales and further information, contact:

Marie Rushton
Business Manager, Publishing and Information Services
Pira International
Randalls Road
Leatherhead
Surrey KT22 7RU

Tel: (+44) 0372 376161
Fax: (+44) 0372 377526

Contents

Chapter 4 Colour reproduction in practice 66

Chapter 5 The colour scanner 101

Chapter 6 Image assembly for colour reproduction 120

Chapter 9 *Specifications for reproduction and platemaking* 179

Appendix 1 *Light sources* 205

Acknowledgements

This book is made up of the work of many of Pira's consultants working in the areas appropriate to the subject. Many of the original words come from course notes, technical reports, and seminar papers. The work leading to these writings has been the work of those consultants; undertaken for the purposes of research projects, of particular or general studies, and as a result of private insights. These have been edited to remove unnecessary material, to update the information, and to give a 'flow' where that has been necessary to the book format. It would not be possible to acknowledge all the sources of this work but the contribution of the following Pira workers is particularly mentioned:

Bardeloe A
Birkenshaw J W
Blue A
Blunden B W
Eamer M
Hancock M G
Hunter Craig T
Johnson A J
Poulter S
Scott-Taggart M
Sunderland B H W
Tritton K T

Chapter 1 *The print processes – an overview*

Printing is the reproduction of images in ink on paper, board or other printing stock. Originals for reproduction can be photographs, artist's work and type matter.

The common denominator of all printing processes is a printing image carrier, *ie* a surface on which the image areas and non-image areas are clearly defined. Image carriers are the plate, cylinder or screen for the different processes.

The three main steps of all printing processes are:

The preparation of the image carrier on which the image areas are defined from the non-image areas.

The application of ink to the image area.

The transfer of the inked image to paper under pressure.

There are five major printing processes – letterpress, offset lithography, gravure, flexography, screen process – and a number of minor processes. The basic differences between the processes are in the way the image and non-image areas are defined on the image carrier.

In letterpress the image areas are raised above the level of the non-image areas – it is a relief process.

In offset lithography the image and non-image areas are defined by chemical and physical properties, and are on the same plane – it is a planographic process.

In gravure the image areas are recessed – it is an intaglio process.

In flexography the image areas are in relief, as in letterpress, but the plates are flexible with 'give' in them.

In the screen process the image areas are open areas in a mesh – it is a stencil process.

1

The image carrier, method of transferring the ink and the type of ink required, characterise the appearance of printed products from each process. The material to be printed – the substrate – may demand one process or another.

Letterpress

The original printing process, it uses metal blocks for illustrations or metal type for text. The process now uses hard plastics plates made by a photochemical process. Thick paste inks are applied to the printing surface by roller application. The process can print directly to the substrate, or by offsetting.

Offset lithography

The litho plate is typically a thin aluminium sheet upon which the image is water repelling (hydrophobic) and the non-image areas are the aluminium which is water receptive (hydrophilic). In use the plate is moistened by dampening rollers, which wet the non-image areas. The ink, similar to letterpress ink, is then applied to the surface by rollers but only transfers to the image areas.

The image is usually transferred to a rubber blanket and it is this which transfers the ink to the paper surface, hence 'offset' lithography. This allows a much enhanced image to be printed, especially on poor quality papers.

Gravure

In gravure the image is usually etched or engraved into the surface of a metal cylinder, although plastic plates may also be used. The cylinder surface is flooded with liquid solvent or water-based ink which fills the engraved areas, the surface (non-image areas) being scraped clean of ink with a doctor blade. The engraved areas are formed by a series of tiny cells, the walls of which act as a support for the doctor blade preventing it from dipping into the engraved image areas and removing ink from them.

Flexography

In flexography the image areas are raised, as in letterpress, but the plates are made flexible and soft. Liquid inks much like those in gravure are used. The printing surface is inked by an 'Anilox' roller which is a cellular engraved roller flooded with ink, and doctored to ensure a constant ink film thickness.

Screen process

The image carrier in this process is a synthetic fabric or metal mesh stretched on a frame. The pores of the mesh are blocked up in the non-image areas and left open in the image areas. Ink is 'squeegeed' across the screen, passing through in the image areas to the printing substrate underneath.

Characteristics

Each of the processes is suited to particular kinds of work, but there is some overlap. Important considerations are:

The time and cost of producing the image carrier.

The type of ink required.

The type of substrate to be printed.

The time needed to make the press ready for the print run (make-ready time).

The possibility of in-line printing and converting processes.

Sheet or reel (web) material supply, or formed product.

The image definition required.

Other processes

We have indicated that there are five main printing processes and a few other, minor, printing processes used in special applications. We shall consider those here.

A form of *intaglio*, or recess, printing is employed for the printing of banknotes. Increasingly banknotes are printed by more than one process, but intaglio is still one of them. The image carrier is an engraved steel, copper or brass plate or cylinder. The image carrier may be produced by hand or computer controlled engraving, by chemical etching, or by a combination of these. The printing of fine tapering lines and a very thick ink film are charac-

3

teristics of the process. The ink is usually dried by infra-red heaters.

Collotype is a process which was used in fine art reproduction but is very rare now. It produces a continuous tone image from a gelatine-coated, originally glass but usually aluminium, printing plate. It is a very difficult process to control and the use of gelatine as the printing surface makes it suitable only for short run work.

Screenless litho was used during the 1970s and relied on relatively conventional materials, but used the grain of the anodised aluminium printing surface to effect a kind of 'semi' continuous tone.

Waterless litho is a modern implementation of the litho process and relies on modern silicon-based non-image areas to reject the ink. The absence of water causes lubrication and cooling problems, but it gives promising results whilst it works. Development continues.

Chromo-litho (literally 'colour stone') is a process made famous by Toulouse Lautrec. It consists of the drawing or painting of a colour separated image, wrong reading, on to the surface of a flat piece of Kalhein stone, a form of limestone. The stone is mounted on a flat bed press and moistened before the ink is applied by a hand held roller. The paper is then placed on the stone surface and pressure applied to effect the transfer of ink to the paper. Many artists still work in the medium, which gives a limited number of impressions. The stone can be re-used by planing the surface and redrawing. The drawing materials are grease-based (oleophilic) and the non-image areas are treated with gum arabic to maintain a water retentive (hydrophilic) surface. In the early part of this century, much commercial printing was undertaken by this process although the more convenient zinc metal plate was often used.

Hot foil stamping is a process which produces very colourful spot (self coloured) colours, including metallics. It is achieved by a relief image carrier, like letterpress, which is placed under great pressure with a foil colorant and the receiving surface in a sandwich. The plate is heated and the colorant transfers to the receiving substrate. Many in-store print departments use the process for the printing of price tickets and displays. Book covers may also be stamped where the result is sometimes referred to as 'gold blocking'. The printing surface can be made up from moveable type like the classic letterpress process but relief metal plates are more commonly used.

Sublimation is a process used for the printing of textiles and consists of a printed transfer carrier, the image of which is transferred under a steam press. It is therefore a two stage process. The transfer is first printed by litho or screen process on to the carrier in wrong reading orientation. The carrier is then applied face down on the receiving material, and steam and pressure are applied. The dye colorant vaporises under the influence of the steam and condenses in the body of the receiving textile. The printed effect is that of a dyed image rather than surface ink.

Ink jet printing is one of the non-impact processes which is useful for 'last minute' additions such as date coding on food packages, or for variable information such as postal addresses on magazine coverings. A print head is mounted on a production line so that products passing under it are sprayed with a controlled dose of ink from a series of fine nozzles. The nozzles are controlled typically by a computer which contains the required data. The surface of the product does not have to be flat as the computer can be programmed to distort the image. Recent developments allow the process to print colour.

Xerographic printing uses an electrostatic image which attracts finely ground, carbon-based toners and transfers them to paper by electrostatic or direct physical means. The toner powder is often fixed on the paper surface by heating. Many photocopiers work on this principle. A recent development has been the introduction of an electrostatic print unit to a standard litho press to achieve 'personalised' print. The image is renewed or replaced after each impression and the image writing computer can be programmed with a whole series of differing images such as addresses. The process can also print colour.

Other substrates

The printing processes are not confined to the usual ink on paper products that we are all used to seeing. Much of the electronics industry relies on several printing processes for its production. Integrated circuit chips are manufactured by lithographic means, and reliance is placed on photographic methods akin to our own platemaking processes. The term 'printed circuit board' means just that. The pattern of conductors on these boards is formed by printing, by litho or screen process, an acid resisting ink on to the copper-coated side of a resin or glass fibre board. The whole is then placed in an etching bath where the non-printed areas are etched away, leaving the copper conductors on the surface of the non-conducting board.

Tin plate printing is an old established process. Litho printing is employed (this was the beginnings of offset litho) to transfer an image to the surface of a flat plate of tin-coated steel. The image may well be distorted in order that when the tin is formed and folded into a box or bowl, the image makes sense in the finished product. Offset letterpress, flexo and screen processes are used to print pre-formed drinks cans and other containers. Edible inks are currently under development for the ink jet printing of 'consume by' dates directly on to foodstuffs.

We should now consider the processes leading up to the actual press run, including the preparation of the image carrier.

The pre-press preparation stages

Most of the processes require photographic film from which the image carrier is made. Hence the preference in some areas for terms like photo-lithography and photo-gravure, or photo-engraving for letterpress. Positive or negative film may be required depending on the process. The filmwork is very similar in all processes if it is only required to print line work such as text and solid colours. The differences in filmwork for each process are apparent when a continuous tone image has to be reproduced.

In litho and relief processes, the ink is applied by roller and it is not possible to vary the amount of ink applied in discrete areas. The image is therefore rendered in a discontinuous or halftone form. The illusion of tone is created by breaking the image up into a series of dots of varying size. If the dots are small, the unit area of print will have proportionately more white paper unprinted, so the tone will appear lighter than when the dots are larger and a greater area of the paper is covered by ink. The halftone nature of the image is produced photographically by the process of screening, projecting the image through a halftone screen placed before a photographic film. Modern scanners generate the halftone images electronically.

The dots produced should not be discernible to the observer and should therefore be small and finely distributed. The fineness of the halftone pattern is described in lines per centimetre (lpcm) or, more commonly in the English speaking parts of the world, in lines per inch (lpi). A typical value for the screen ruling is 60lpcm for general litho, but depends on the quality of the paper used. Newsprint may be rendered in 30lpcm.

In photogravure the film is required to be continuous tone, the screening into cells taking place at the cylinder surface during cylinder manufacture. The cells are of constant area but vary in their depth depending on the density of the continuous tone positive. The use of litho separations for gravure is almost universal now, although a gravure screen is used as well as the litho screen to produce cell walls. Increasingly, gravure cylinders are produced electro-mechanically. Scanning methods are used either direct from the original or from the litho halftone films. The scanned data is passed to a large lathe-like engraving machine which cuts the correct size cells into the surface of the printing cylinder. The cutter on these machines produces a cell which varies in both depth and area.

Colour reproduction

To reproduce a coloured original which may be anything from a colour transparency to an oil painting, it is first necessary to separate the original into records of the amount of primary colours contained in each point of the original. This is achieved by photographing the original through colour filters. It may be done using a graphic arts camera, by contact printing, or electronically using a scanner. Over 95% of colour separations made worldwide are produced by scanning. We will, therefore, concentrate on that method, although some consideration should be given to photographic colour separations as these provide a useful model of the principles at work in a scanner.

The result of colour separation is a set of films, usually four, each representing the amount of a particular primary colour of the printing process. The films will ultimately be printed separately using the appropriate coloured ink, either yellow, magenta, cyan, or black.

The yellow printer contributes to all the colours in the print that contain yellow, namely reds, greens, browns and black. Similarly the magenta contributes to the reds, blues, browns and black, and the cyan to the blues, greens, browns and black.

There are problems in the printing processes where the practice does not conform to the theory and these problems have to be addressed by the colour separation process. The major problems, in colour reproduction by all printing processes, are that the inks used are far from perfect in their colour reflectance and absorption, and it is unlikely that a perfect set of inks will ever be produced. In general the imperfections are that the magenta ink is insufficiently blue, and cyan is lacking in both blue and green, although the yellow is nearly ideal. To compensate for colour inadequacies, colour correction has to be carried out during the colour reproduction process so that the appropriate amounts of ink are printed to achieve an acceptable colour reproduction.

There are three main methods of achieving colour correction: hand retouching, photographic masking, and electronic methods accomplished during scanning. In each case the principle is the same – to compensate for unwanted absorptions in each of the primary ink colours. Since most ink absorption errors are in the blue and green regions of the spectrum, so most of the colour reproduction errors will also be in the blue and green areas of the printed result.

Corrections can be made at various stages of reproduction, depending on the printing process. Hand correction is too laborious and expensive for all but the most straightforward or exacting work. Correction in all processes is more efficiently carried out by photographic masking methods or using electronic scanners.

Hand retouching

Retouch artists chemically alter the density of the colour separation films to change the amount of coloured ink that will be printed. Materials used may be chemical etches, applied globally by bath or locally by brush, intensifiers, or the application of retouching dyes. In litho, once the image carrier has been made no further retouching is possible – although in gravure the metal of the image carrier can be etched. Retouch artists require both skill and experience to operate efficiently. Their learning period is considerable. The advent of colour scanners has largely reduced the need for retouch artists but their counterparts, colour retouchers, are still employed in the transparency and photographic retouching fields.

Photographic masking

Photo-mechanical correction is a selective alteration of the photographic densities using low contrast images which are either positive or negative, usually unsharp, masks.

When making colour separation negatives from the original, by the interposition of a negative mask exposed through a different colour filter, it is possible to increase the strength of the 'worked' colour. At its zenith, this method used dye image, coloured masks which were kept in place during the separation process, and applied the correct modification to each colour separation as it was exposed through the colour separation filter.

It is possible to make 'straight' colour separation negatives and apply photographic masks to the making of positives, where the process demands them. Various methods of masking were developed but the most successful were single- and double-overlay methods. In these methods, very specific corrections could be made and so they lent themselves to the production of multi-colour (*ie* more colours than four) printing.

Electronic methods

Colour scanners correct tone and colour densities automatically by means of a pre-programmed computer. The original is scanned by a light which, in turn, is received by photocells. Fluctuations of the current from the photocells, corresponding to the red, green and blue densities of the original, provide the input signal to the computer. This estimates the correct amount of yellow, magenta, cyan and black ink needed to reproduce the scanned area.

Film assembly and planning techniques

The colour separation process provides, typically, four films for each colour picture. Film is also made from any line work, text and tint work required on the finished job. The films have to be combined to make the plate-ready films, where all elements of the job are in their correct positions. Achieving this is the task of the next process in the chain – film assembly.

Planning considerations

For most types of work, before any planning takes place, it is necessary for a layout to be drawn. The layout should show the position of all trims and folds and indicate where the various images should appear. The planner needs to be aware of the sheet size, grain direction, plate size, plate clamping allowance, paper grip allowance, folding arrangements and the graphic details of the job. A knowledge of specific colour reproduction problems is also necessary. The rendering of a large black area, for instance, may be achieved in a number of ways. Simply printing a solid black area usually results in a low density black. It requires the addition of other colours to render a sufficiently dark black. This may be achieved by printing a 60% tint of cyan under the black or it may require the addition of a grey printed in three colours. The planner has to select the best way for the job, paper and printing process in use.

Preparing the layout

The layout will be used to determine all the topographic details of the finished job. It is a sheet of dimensionally stable material upon which is ruled the page sizes and positions, the grip allowances, the centre line of the printed sheet, and other elements such as any bleed allowance, gutters and so forth. All measurements on the layout are made from the centre line or the base line to preclude the occurrence of accumulating errors. The layout may be referred to during film assembly purely for the information it contains or, more usually, it will be made on transparent plastic sheeting, called foil, and used as part of the assembly as a visual reference for the position of the pictures and text.

It is important that the assembled films are accurately placed on the image carrier, and the layout will contain either register pins or alignment marks to achieve this.

Film planning

The layout is mounted on a light table so that the markings show up. A new, clean foil (or 'planning flat') is placed over the layout and one common colour separation film (usually cyan) from each picture is positioned according to the layout marks. Each film is held by adhesive tape to the foil. When all the graphic elements, including text, tints and pictures, for the cyan are in place, the layout can be removed. The other colour foils are assembled by reference to the cyan foil to ensure good register.

It is often at the planning stage that elements like solid colour tint blocks are introduced. These can be produced by assembling areas of photo-opaque material to the foil. Film tints are also cut to size and assembled. All cut edges of film and adhesive tape will be rendered on the printing surface as though they were solid lines, so the film planner has to use some method of avoiding or removing these. Any dust, scratches and unwanted marks on the foils will also reproduce.

The most efficient means of ensuring that a clean plate will be made is to use a separate foil, known as a 'burn-out' mask, to protect all the printed areas of the plate. The plate is then exposed first with the burn-out mask and, in a second exposure, with the assembled colour foil. The mask may contain a number of graphic elements itself – text and logos which are to be tinted, for example. The burn-out mask then leaves a shape on the plate, into which, by the second exposure, the tints are placed. The use of the burn-out mask also allows the picture edges to be rendered in a clean, sharp boundary.

The film assembly described above is common to most colour printed work, but may be varied in detail for the demands of each job. If negative plates are in use, the method of planning subsequent foils visually to the first is difficult because of the opacity of the negatives. In this case the first colour planned may be photographically copied on to a separate light sensitive foil, such as Ozalid material, to produce a positive transparent, coloured dye image. This is then used as the reference image for assembling the succeeding colour foils. This method is known as the colour key process.

Film planning is increasingly being replaced by electronic assembly. The images, which have been scanned, are stored by a computer and placed in their finished position on a video screen image of the page. The electronic page composition system, as it is called, then produces finished film colour separations with all graphic elements in place.

Making the image carrier
Litho platemaking
Litho is the most commonly used printing process worldwide. We shall, therefore, consider litho platemaking first, then consider the other processes by comparing them to litho.

The litho printing plate is supplied as a thin sheet of reasonably pure aluminium normally ready coated with a light sensitive coating ('surface presensitised'). It may be positive or negative working to choice, which means that the coating will either harden or decompose under the influence of actinic (ie short wave, ultra-violet) light. The plate is usually made by contact exposure to the assembled foils, one plate for each of the colours to be printed.

Contact exposure is achieved by laying the assembled foil on the coated surface of the plate and putting the couplet in a vacuum contact frame. Since the foil is transparent but the image on it is opaque (or *vice versa* for a negative), any light can only effect the coating in the transparent areas. A powerful ultra-violet lamp is then used to expose the plate through the foil. Chemical processing washes the coating from the non-image areas of the plate, leaving the image in coating on the plate. One of the properties of the coating is that it is hydrophobic, so able to repel water and attract and retain oil-based ink.

The non-image area, from which the coating has been removed, is the aluminium of which the plate is made. This surface is typically grained and anodised in manufacture. The anodic surface is hydrophilic so able to attract and retain water. The grained surface helps in the retention of water.

Letterpress plates or blocks
Letterpress is the oldest of the printing processes, originating in China during the 8th and 9th Centuries when wood blocks, cut with relief images, were used. The relief areas are inked by a roller so that only the raised surface holds the ink. The printing substrate, usually paper, is then brought into contact with the inked surface and pressure applied. The ink requires sufficient tack to stick first to the printing surface, then to the paper, in order to transfer.

A number of methods have been used to produce the printing surface in the past, carving and later moulding being the most prominent before the 20th Century. During this century, chemically etched zinc and copper have been favoured until recently when plastic has taken over. Both rubber and plastics plates have been made by moulding from original metal plates, but the recent developments have been in the direction of making the plate in much the same way as litho plates, by exposure to actinic light and chemical processing.

The plates are usually supplied as plastic sheets about 2mm thick, often with a metal backing, and are photosensitive, *ie* they will harden when exposed to light. These are known as photopolymer plates. The metal backing may be steel so that the plates can be mounted on the printing cylinder by magnetic means.

The exposure, through a negative, is as described for litho plates above. The chemical processing includes the washing out of non-image areas so leaving the image in relief. The plates are then washed and dried, when they are ready for the press. Automated processors are used in high production installations such as newspapers.

13

Gravure cylinders

A steel cylinder is copper-plated by electrolytic deposition. The copper traditionally is chemically etched to produce the image. Exposure is made to a gelatine sheet, in the flat, first to a continuous tone positive, then to a screen which will form the cell walls. The gelatine sheet, called carbon tissue, is transferred to the cylinder by rolling it into contact while feeding the surface with water. The gelatine sheet adheres to the copper where, after development, it leaves a relief image of hardened gelatine to form an etch resist of varying thickness. The etch, which works most readily on those areas with the thinnest covering, is then applied, etching the copper to a depth dependent on the density of the continuous tone positive.

An otherwise similar system developed in the 1970s was called litho-gravure, where the continuous tone positive was replaced with a litho-style halftone separation. This substantially reduces the costs of cylinder production and enables a litho proof to be made of the films before cylinder making.

Cylinders can also be made by electronic engraving. On the Helio-Klischograph, as the machine is called, cells are produced by engraving with a pyramid-shaped diamond stylus. For deeper cells, the stylus penetrates further into the copper, so that the area of the top of the cell also becomes greater. These machines, whilst being very costly, reduce the cylinder making time to about two hours. Chromium plating is used to extend the life of both etched and engraved cylinders and they can be re-plated for very long runs.

Flexographic plates

Flexo plates may be of rubber, polymerised plastics or laser engraved rubber plates or cylinders. The rubber plate may be moulded from a metal master, such as a letterpress plate, and thermosetting plastic compounds are used for this.

More commonly used, these days, are polymer plates, which carry light sensitive additives. These are usually referred to as photopolymer plates, similar to those used in letterpress, and are made directly from film negatives. The plates may be supplied as backing sheets of plastic or metal, and coated with a liquid polymer prior to exposure, or alternatively supplied as a pre-coated plate. Exposure to ultra-violet light, through a negative, follows, which hardens the photopolymer in the image areas. The non-image areas are then washed away, using solvents, water or air, leaving the image in relief. The substance of the plate is flexible and soft, whereas letterpress photopolymers are hard.

Rubber suitable for the flexo process can be engraved, using a powerful laser, on a scanning machine. On the machine, separate plates may be mounted and engraved, or a flexo roller can be produced. A computer controlling the laser may be programmed so as to produce 'seamless' images, *eg* for wallpaper printing. The machine can take the cutting information either from an onboard input scanner, or from electronic image data through the computer.

Screen printing

The image carrier is a taut screen carrying a stencil. The screen can be made of synthetic fibre or metal. It was originally called silk-screen because silk was used to make the screen. The pores of the screen mesh are obscured in the non-image areas and open in the image areas. The ink is forced through the open areas of the screen by a squeegee to print. The printing pressure is very low and the process therefore lends itself to the printing of fragile objects or awkward shapes. Toys, instrument legends, domestic appliances, and even vehicles are screen printed.

The screens are now mostly made from synthetic monofilaments such as nylon or polyester, the latter where dimensional stability is important. Metal screens, which are much more expensive, are used where a long life and complete dimensional stability are required, such as with printed circuit boards. The screen material is stretched and mounted on a frame. The correct tension is important to avoid misregister, premature wear of the stencil or splitting of the screen. The screen is degreased before the stencil is applied.

The stencil may be hand-cut from a laminated film material and stuck to the screen. This is the common method for printing designs with large areas of colour of simple geometric shape. The stencil can also be formed photographically. A number of successive coatings (to ensure complete coverage) of dichromated resin may be applied to obscure the whole surface of the screen. Prior to exposure, the coating is water-soluble but upon exposure to light it hardens. If exposure is made through a photographic positive, the coating is hardened in the non-image areas. The image areas are then washed away leaving the screen mesh open. This is the direct stencil. Any openings in the non-image areas of the screen can be seen by viewing the screen on a light table, and filled using an opaque filler. Stencils may also be made indirectly by exposing supplied photosensitive material and transferring it to the screen after exposure.

The printing processes compared

It is not intended here to describe the processes further; each of the processes deserves a book of its own. It is useful, however, to summarise the kinds of work undertaken by each process, its advantages and its limitations, and these are shown in Table 1.1.

Table 1.1

The printing processes compared

OFFSET LITHO

Typical work
Newspapers
Magazines
Stationery and forms
Promotional material
Books (illustrated)
Posters and maps
Directories and catalogues

Advantages
Good illustrations on most papers
Fine screen rulings possible
Low cost plates
Many plate types
Short lead times
Simultaneous two-sided printing (perfecting) possible
Wide range of sizes and configurations of press

Limiting factors
High print waste
Only oil-based inks
Poor sharpness of print compared with letterpress
Limited colour saturation
Places high demands on paper
Limited plate life

LETTERPRESS

Typical work
Security print
Newspapers
Roll labels
Books (text)
Packaging (paper and board)
Pre-formed cans and other containers

Advantages
Legibility of text
Water-miscible inks possible

Limiting factors
Coarse screen ruling compared with litho
Poor illustrations on uncoated paper
Variable pressure distribution
Plate image control difficult compared with litho

17

FLEXOGRAPHY

Typical work	Packaging (film)
	Roll labels
	Bags and envelopes
	Cartons
	Corrugated cases
	Newspapers
	Wallpaper
Advantages	Suitable for non-absorbent substrates
	Lower press and origination costs than gravure
	Simple inking system
	Continuous images possible
	Variable cut-off
Limiting factors	Coarse screen ruling
	Difficult to achieve a vignette
	Limited fine detail

GRAVURE

Typical work	Magazines
	Catalogues
	Packaging (film)
	Packaging (board)
	Wallpaper
Advantages	High colour saturation
	Suitable for non-absorbent substrates
	Low print waste
	Ideal for wide tone variation illustrations
	High production speeds
	Continuous images possible
Limiting factors	High cost image carrier (cylinder)
	Long lead times
	Storage and handling of cylinders difficult
	Fine lines and text show screen pattern
	Requires comparatively smooth paper
	Mottled solids

18

SCREEN PRINTING

Typical work

Finished product printing
Point of sale material
Posters
Fabrics and transfers
Non-uniform fragile surfaces
Electronic circuit boards
Signs

Advantages

Low cost press
Low cost image carrier
Suitable for short runs
Very intense colour
Very thick ink films
Can print on almost any material
Can print unusual inks and coatings

Limiting factors

Limited fine detail
Slow production speeds
Coarse screen rulings

Chapter 2 *Fundamental principles*

This chapter deals with the fundamental principles necessary for colour reproduction in printing. The nature of light is an important prerequisite to an understanding of how colour reproduction is achieved, why it doesn't work as well as we might hope, and what compromises can be made to optimise the process. The halftone method is explained, along with its implementation, since nearly all photo-realistic printing processes rely on it. Densitometry is widely used in the printing processes as the major controlling measurement. We shall therefore examine densitometry also.

These three major topic areas are essential to the understanding of the remainder of the book, which assumes that understanding. Readers who are conversant with colour, halftone and densitometry are advised to check the contents of this chapter for the purposes of revision.

Colour

Colour is an experience enjoyed by almost all the human race from a very early age. As we grow we learn to recognise, and often name, specific colours such as 'sky blue' and 'grass green' and yet we can never be sure that everyone derives precisely the same sensation from a given stimulus. There are an important minority with colour vision defects who obviously do not, but we generally expect that the remainder do. All we can really assert is that the sensation received is similar.

When we look at the physical causes of colour it is apparent that these may be defined easily, and therefore any personal differences of sensation are very much a problem of perception. In order to gain a full appreciation of colour theory, and the problems of colour matching, it is essential to consider perception in some detail. Some background knowledge is required prior to such consideration.

The sensation of colour is achieved when electro-magnetic waves between the limits of approximately 380 and 760 nanometres are incident upon the eye. These particular electro-magnetic waves are commonly referred to as light waves. The shorter wavelengths produce the sensation of violet and as the wavelength increases the sensation changes to blue, green, yellow, orange and red successively. Thus whenever a visual sensation is received which is red, for example, it means that the light falling on the eye has a predominance of long waves.

The physical nature of colour
Saying that light which appears red contains a predominance of long waves to some extent begs a question. In the real world, the objects around us appear to have a variety of colours and yet the sources of light energy, such as the sun, frequently appear to have little or no colour. What is it then that gives rise to the appearance of colour in objects? The answer lies in the demonstrations of Newton that 'white' light contains all the wavelengths within the visible spectrum and the fact that when light falls on objects in the real world it is modified by them.

For observers to see an object it is necessary for them to receive a complex pattern of light rays which have come from that object and contain the information required. Yet the majority of objects do not in themselves emit light, as can be demonstrated easily by removing any sources of light energy. It is quite obvious, therefore, that the light necessary for perception generally comes from an outside source of energy and it is the way in which this is altered by the object that determines how the object is seen.

When electro-magnetic waves fall on any object they can do so in only two ways: they may be transmitted through, or reflected by the surface of the object. At the surface the light waves will be attenuated. They may be absorbed, refracted or diffracted, depending on the nature of the surface. Since the incident light is composed of all the spectral colours (if it is white), and since each colour is distinguished by its wavelength, it may be that different colours (wavelengths) are attenuated by differing amounts.

The surface of an object which appears red when viewed in daylight has generally attenuated a predominance of the shorter wavelengths of the visible spectrum and reflected or transmitted the remaining light to the eye of the observer. The light energy absorbed at the surface is usually re-emitted as heat. Thus it is apparent that perception of colour in objects is totally dependent upon the way in which that object attenuates the light waves incident upon it. The relative attenuations of the various wavelengths determine its colour and the total attenuation determines the lightness of that colour. The manner in which the light is reflected or transmitted is important in determining gloss, texture and translucency.

The perception of colour
The subject of colour perception is an area of science which is still a long way from being completely understood. Quite apart from the lack of complete understanding of the mechanics of the visual process there are also problems of adaptation and contrast effects.

A great deal has been written and discussed concerning these aspects and in this work we can do little more than scratch the surface, but they probably remain as the most significant technical stumbling block to a complete automation of colour matching procedures or quantitative assessment of colour reproduction.

For many purposes the mechanism of the visual process or the influence of adaptation effects may be ignored and then colour measurement will offer a perfect definition of a colour, but, as we shall see, this is all too frequently not possible.

For very many years the theory of colour vision most widely accepted by colour scientists was that first postulated by Young in 1807 and later developed by Helmholtz in 1852 and 1866.

It was suggested that in the retina of the eye there exist three receptors, one principally responsive to short wavelengths of the visible electro-magnetic spectrum (blue), one to middle wavelengths (green) and one to long wavelengths (red). These receptors, known as cones, are connected to the visual cortex of the brain by a series of neural networks and together these produce the colour perception.

This particular theory has been supported by the evidence which showed that all colours may be matched with mixtures of three, widely separated, monochromatic (single colour) radiations (see Plate 1). There is little doubt today that this is the accepted hypothesis of the mechanism of the response of the eye to colour, despite the fact that no cone pigments have, as yet, been isolated definitely. Nevertheless, it does have certain shortcomings as a complete theory of colour vision, partly in the classification of those with defective colour vision, but perhaps of more interest, in the problem of the perception of yellow.

Whilst there is no doubt that yellow light may be produced by suitable mixtures of red and green light, at the same time it is a colour which has no apparent redness or greenness about it. This type of property is shared by only three other colours; red, green and blue. These four colours together are known as the psycho-logical primaries and have unitary hues, ie each of them is per-ceived as being totally independent of the other three primaries. Based on this aspect of colour vision, Hering in 1878 developed an 'opponent response' theory. He postulated the existence of three pairs of response processes – white/black, yellow/blue and red/green – which take place in the visual mechanism, and suggested that these explain the existence of the psychological primaries.

As a complete theory of colour vision, the Hering hypothesis does not hold, simply because of the colour matching evidence de-scribed earlier. Nevertheless, it can be developed to a form which explains fully the phenomenon of defective colour vision.

It is likely, therefore, that any hypothesis of colour vision which is finally proved correct is likely to contain within it both the Young-Helmholtz and Hering concepts and define some complex neural connections bridging the gap between them. One such theory, is postulated in 'A model of colour vision for predicting colour appearance' - *Measuring Colour*, R W G Hunt (1987) Chap-ter 8.

Physiological processes in colour perception
The conversion of light energy into nervous energy takes place in the light sensitive retina at the back of the eye and, therefore, it follows that this must contain colour sensitive receptors.

Fig. 2.1 The eye in cross section and showing the retina surface

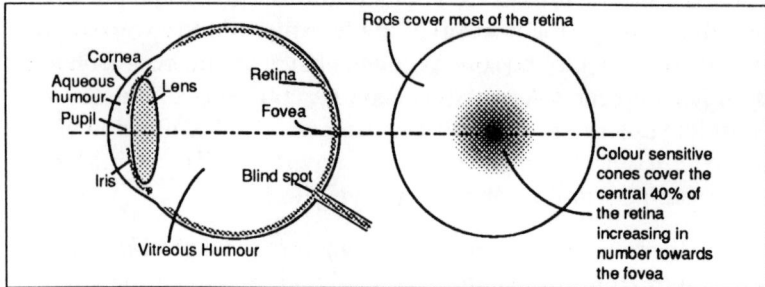

These are called cones and, as described earlier, it is believed that three types exist with differing colour sensitivities. The retina also contains receptors known as rods which are sensitive to light of short wavelengths and operate primarily at low levels of illumination.

Within the receptors is a laminated structure. It could be that this has the photo-chemical substance distributed over it, or it could have some physical purpose analogous to a wave guide so that the receptors tune to the desired wavelength.

The next process to consider is the transmission of information from the retina to the visual cortex of the brain. If we assume there are red, green and blue centres in the visual cortex, it is difficult to see how the impulses get there without getting confused since it is unlikely that the impulses are different in themselves.

Pedler reports an intricate network of fine connections running across the synoptic layers of the eye and providing a vastly greater opportunity for interaction between the signals from individual receptors than had previously been imagined.

Thus we can imagine this region acting like a computer and coding the information in some way. It is also claimed that this is necessary because the number of nerve fibres is very much smaller than the number of receptors, and with the relatively large amounts of light entering the eye, cross-linking is necessary to allow all the information to reach the optic nerve.

We have further synoptic layers at the lateral geniculate nucleus and De Vallois has produced evidence to show that the information from this part of the visual system divides into two general classes, one class apparently responsible for the transmission of intensity information, the other class for colour.

The 'colour' cells respond by excitation or inhibition according to wavelength, and are coupled in opponent pairs by their response to red and green, and to yellow and blue. This suggests that transmission in the visual cortex may be in terms of the Hering opponent colour theory of colour vision.

To relate the spectral composition of a colour to the visual sensation, we find that certain wavelengths give rise to certain colours and if we consider the spectrum we find that the boundaries between colours are very ill-defined. However, we can select a number of definable colours – red, green, yellow, purple, *etc* – each of which we can call a hue.

Thus the hue of a pigment is defined by its similarity to a spectral radiation. Its depth of colour will depend upon its concentration, the lower the concentration, the weaker the colour and, therefore, the less saturated. Finally there is the total amount of light emitting from the colour, which will determine its brightness or lightness, (the former being applied to light sources).

It is not strictly true to say that the perceived brightness or lightness of a colour depends on the amount of light it contains. Two colours may emit the same amount of light, but have different apparent lightness, because of the lightness of their surrounding areas. However, by its hue, saturation and lightness we have a method of determining or specifying a colour.

Adaptation
When we go from light to dark, or *vice versa*, the eye adapts to allow us to see comfortably. This comes about in four different ways:

> *Rod-cone or cone-rod switch.*
>
> *Iris diameter variation.*

Neural adaptation.

The photo-chemical cycle.

Changes in colour sensitivity are less frequently noticed, but they are continually occurring and may be quite large. For example, white paper looks white, regardless of the illuminating source's spectral composition (as long as it covers the visible spectrum).

If it were not for adaptation the white paper would look reddish under tungsten light and bluish under daylight. Nevertheless exact adaptation rarely occurs and with light sources with irregular distributions, such as the fluorescent tube with its spectral lines, the colour distortion may be considerable. Another factor which helps in this colour constancy is the relation to surrounding colours; if the colour of the source changes then all the colours will be distorted, which helps in the recognition of a colour.

Metamerism

Metamerism may be defined quite simply as the property of the eye and brain to receive the same colour sensation from two objects with different spectral energy distributions. In other words, despite the fact that at many wavelengths in the visible spectrum the visible light energy emitted by the two stimuli may be different, both appear the same to the observer. Two objects matching in this way are known as metamers. The reason why this may occur depends upon the fact that the eye has three receptors which are colour sensitive and means that the solo requirement for two colours to match is that the total light energy, with respect to the sensitivity of the receptors, is the same for both objects.

With object metamerism we are concerned with differences between the spectral reflectance of the objects. Consider an example where we have two objects which have quite different spectral reflectances and yet for a given observer under a particular illuminant they match for colour. If we now change either the observer or the illuminant, it is likely that the match will break down and this is the situation which is of most interest. An example of a pair of objects which are metamers when seen under daylight but not under tungsten light is shown in Figure 2.2 and here it is apparent that the biggest difference between them is the spectral reflectance at the long wavelengths.

Fig. 2.2 Luminance of objects which are metamers under daylight but not under tungsten light

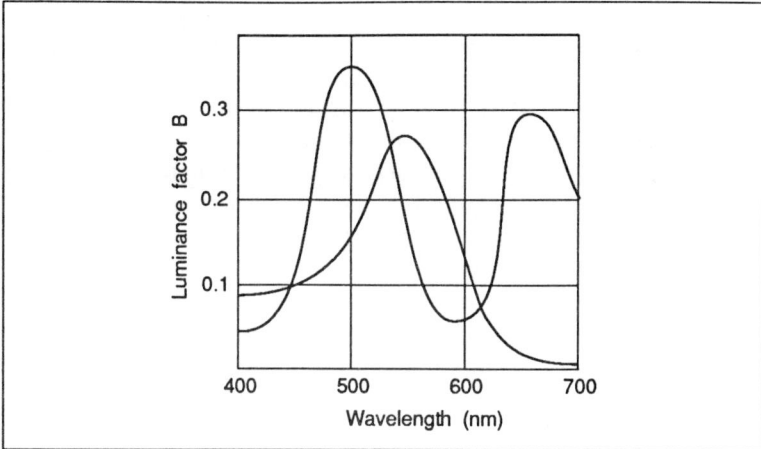

Since the relative light output at the longer wavelengths of tungsten light is very high compared to that of daylight, and furthermore, the sensitivity of the eye at long wavelengths is fairly low, the combination of these effects is sufficient to ensure an approximate match under daylight, despite the large difference in spectral reflectance.

The importance of metamerism rests on the fact that practically all colour reproductions exhibit the effect. Most methods rely on the mixture of three primary colorants to match a colour. In almost every case the resultant match between original and reproduction is metameric. Without this very important property of the eye all methods of colour reproduction in common use would be impossible.

Colour reproduction principles
The fact that the eye has only three types of cone has been largely confirmed over a number of years by a series of experiments designed to lay the foundation stone for colour measurement. It had been known for a number of years that mixing coloured

27

lights would give rise to a whole range of other colours, but it was finally in the mid-1920s that exhaustive tests were carried out to establish whether or not different observers mixed approximately the same amount of each in producing a match to a specific colour. The fact that such a similarity was found to exist showed deviations between observers was small. This led to the possibility of defining a method of colour measurement based solely on the mixture of three different lights, one red, one green and one blue, since it was shown to be possible to produce all colours from those three alone. This principle is fundamental to most modern methods of colour reproduction whether by printing, photography or television.

Additive colour mixture

Additive colour mixture is based on the principle outlined above and is the technique used in television. The video screen consists of an array of red, green and blue phosphors and the various colours displayed on the screen (including white) are obtained by causing the phosphors to emit more or less light depending on the signals received. Thus, as the relative amounts of red, green and blue are altered, so a whole gamut of colours may be produced.

Despite our earlier assertion that all colours may be matched by mixtures of three coloured lights we find that this is not immediately possible for spectral or near-spectral colours, unless we add one of the three matching stimuli to the colour to be matched, and use none of it in the match. That is, the spectral colour is desaturated by addition of one of the matching stimuli and the resultant colour matched by the two remaining stimuli. This is the same as having a negative amount of one of the stimuli in the match and if this is accepted then the ability to match any colour by a mixture of three stimuli is achieved.

The reason for this negative requirement lies in the fact that the sensitivity of the receptors of the eye overlap, as may be seen in Figure 2.3.

Fig. 2.3 Sensitivity of receptors

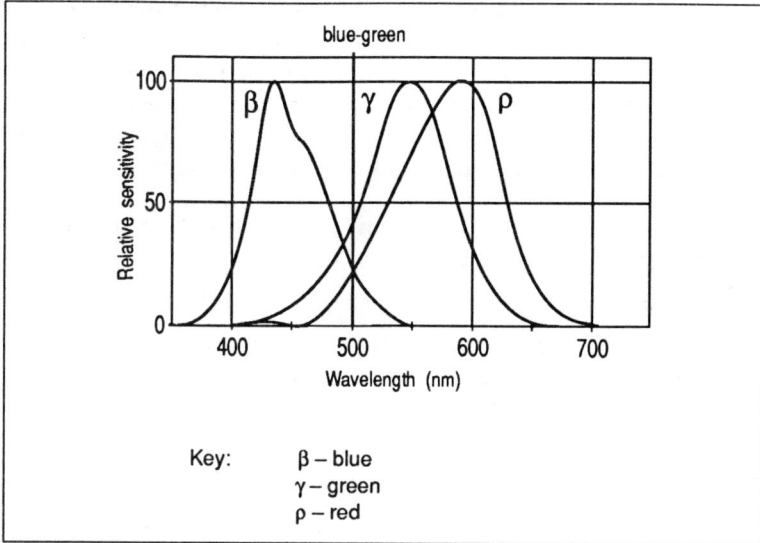

If, for example, we consider a blue-green at 500nm, we can see that as well as stimulating the blue and green receptors, it also stimulates the red.

If we attempt to match this blue-green with a blue of, say, 460nm, and a green of 530nm, we inevitably stimulate the red to a greater degree than with the monochromatic 500nm radiation. This means we must add red to the spectral blue-green to enable us to match it. This gives us a negative amount of red in the mixture curves.

This requirement for negative stimuli is something of a problem in colour reproduction, since in general it means that any colour which requires a negative stimulus cannot be reproduced, and it also gives rise to errors in all other colours which can only be partly compensated for in general.

However, this is a fairly complex problem. Suffice it to say that in principle all colours can be matched by a mixture of these primary coloured lights.

Subtractive colour mixture

Whilst additive methods can be devised for photographic reproductions where the photograph is viewed by transmitted light (*eg*
slide projections) they are not practical for any reflection print
reproduction. Even additive transmission systems suffer from
disadvantages, and in practice most photographic and all printed
reproductions are based on the subtractive principle.

At first sight, subtractive colour mixture appears quite different
from additive methods but as we shall see there is a defined
relationship between the two.

Whether there is any advantage in trying to define a subtractive
system in terms of additive mixture is a debatable point. On the
whole, we shall treat subtractive mixture without reference to
additive methods but it is helpful in understanding the process,
and its limitations, to be aware of the relationship.

We saw earlier that it is possible to produce all colours by
additively mixing red, green and blue light. When the three are
combined in the right proportions white is obtained, and when
no light is present black results.

Red plus green gives yellow, red plus blue gives magenta and
blue plus green gives cyan (see Plate 1). As the proportions of
colours change so all the other hues and levels of saturation are
achieved.

With subtractive colour mixture the principle is different in that
white is the starting point. It is the transparent film in the case of
projected films and unprinted paper in the case of prints. In order
to produce colours, light of particular wavelengths is subtracted
from this white leaving only the wavelengths necessary to
achieve the desired colour impinging upon the eye.

Remembering that the eye has receptors responsive to red, green
and blue light, it is these colours that have to be subtracted and
thus the primary colours for a subtractive system are simply
those that absorb (subtract) red, green and blue. These are cyan,
magenta and yellow, the three primary colours used in printing.
Cyan absorbs red, magenta absorbs green, and yellow absorbs

blue. Thus if the three are added, red, green and blue are absorbed and this results in black (see Plate 2).

Cyan plus magenta absorbs red and green leaving blue, magenta plus yellow absorbs green and blue leaving red, and cyan plus yellow absorbs red and blue leaving green.

As this description stands we have only produced eight colours and obviously this is not sufficient. Suppose, for example, we want to produce a saturated orange. Our additive mixture tells us that an orange is produced by mixing red and green but such that the proportion of red is greater than green. Depending on whether we require a yellow-orange or a red-orange, so the proportion required varies.

With subtractive mixture, therefore, to obtain a saturated orange we need more red light than green but no blue. This tells us that we need a lot of yellow to subtract the blue, less magenta to subtract some green and very little cyan to subtract little of the red.

Similarly to obtain other hues between the primary and secondary colours the same sort of argument will apply. Thus the next difficulty is to find a method of varying the amount of cyan, yellow and magenta applied.

In photographic reproductions this variation is achieved by varying the concentration of the dye layer. This is shown in Figure 2.4 and it can be seen from this that as the concentration of dye increases so the amount of light transmitted is reduced.

Fig. 2.4 Variation of colorant applied in photographic reproductions

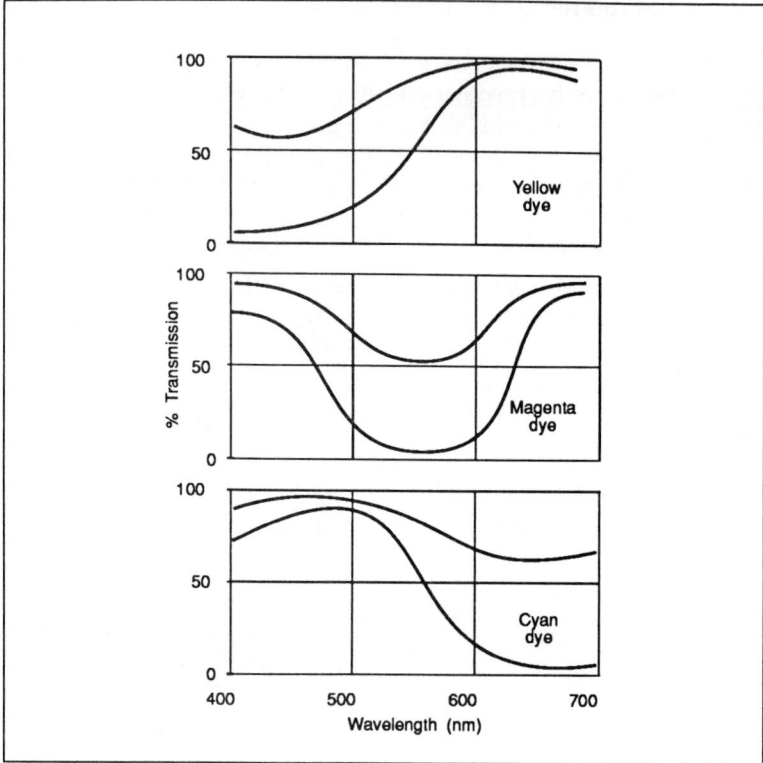

Fig. 2.4 Variation of colorant applied in photographic reproductions

Whilst transmission varies throughout the wavelength range it is most pronounced in either the red, green or blue depending upon the colorant involved and thus achieves our purpose of variation in amount of red, green or blue light.

With the printing processes it is not generally possible to vary the concentration (or film thickness) to achieve modulation of the red, green and blue absorption. The primary exception to this is conventional gravure where the amount of colorant printed depends upon the depth of etch and hence modulation is achieved by variation in film thickness. All other processes, however, do not have this flexibility, and in letterpress, lithography, flexography and invert halftone gravure, the amount of colorant is varied by introducing the halftone principle. In order to obtain the modulation required, a 'dot' of ink is printed and the size of this depends upon the amount of red, green or blue reflectance required.

Principles of halftone reproduction

When reproducing a photograph, using conventional commercial printing processes, we are presented with a major problem. Apart from gravure, none of them is able to print a light rendering of the ink. When using just a black ink, for example, printing image carriers put a solid layer of ink of even thickness on to the paper. It is a totally 'on or off' situation. Printing will either lay down a single thickness, and therefore single darkness, of ink, or leave plain white paper. As the picture information is contained in varying darknesses, or greys, without some simulation of different greys the processes would not be acceptable. The halftone process is a means of simulating tonal greys with a single black ink.

A monochrome continuous tone image is one in which the tones are differentiated by shades of homogeneous grey. All the information about a subject which has been photographically recorded is held in changes of grey, from the pure white of the base paper, through to the maximum black that the material is able to produce. With an original photograph the transition from white to black does not occur in regular steps but is continuous. It is a gradual darkening, through all the various greys which make up the total image. This is why a photograph is referred to as a continuous tone original. For colour the same idea applies but to all the intermediate colours including greys.

An original drawing made by an artist is created by varying the quantities of pencil or paint on the base material. More pencil equals darker image, less pencil equals lighter. The tonal differences are continuous in direct relation to the quantity of the artists' material applied.

Visual integration
The human eye has definite limitations in accurately seeing very small detail. From a normal reading distance, it is unlikely that one would be able to distinguish as separate, individual lines

drawn at 0.20mm apart or closer. The eye will integrate, or join together, the lines. Because there is white paper between those lines, which is also integrated into the total image, one would not see the result as solid black, but black diluted with white. In other words a grey. Increase the quantity or ratio of white to black and, when integrated, the grey will appear lighter. The same effect can be achieved using dots of solid black. By varying the size of the dots relative to the amount of white background a full range of integrated greys are achievable.

Without magnification, the eye is unable to see detail finer than 0.1mm. The finest pattern that can be resolved would be five line pairs (with one black line and one white line per pair) per millimetre.

A television picture is constructed of 625 horizontal scan lines. When viewed from a normal viewing distance the scan lines are not visible, only the total picture is seen. The eye is integrating the lines into an apparently continuous tone picture.

Halftone
Ways of visually simulating a range of tonal greys, with only a single density of black, have been understood by artists for a long time. Engraving and etching lines of varying thickness and frequency (see Figure 2.5) are techniques of illustration which go back beyond the start of our industry. However these rely on an artist's skill and dexterity. The end result is only one person's impression of the scene or object. Our industry demands a very lifelike copy of the scene we want to reproduce, as well as a production technique which is both fast and repeatable. Both these demands make the artist's simulations unacceptable.

The printing industry uses a photo-mechanical process to achieve a similar end result. It does however still rely on the eye's inability to resolve fine detail, and the brain's trick of integrating different areas of black and white into grey.

Translating the range of greys from the continuous tone original into an image consisting solely of solid dots of different sizes (see Figure 2.6) is known as halftone conversion.

Fig. 2.5 Engraving (enlarged)

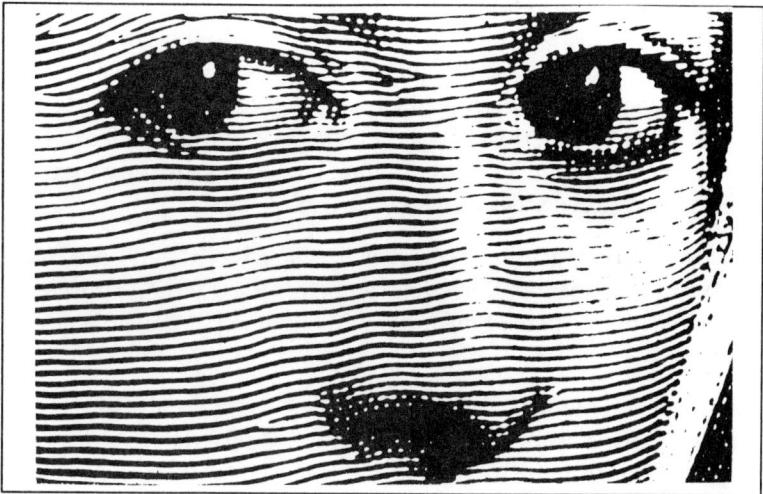

Fig. 2.6 Halftone (enlarged)

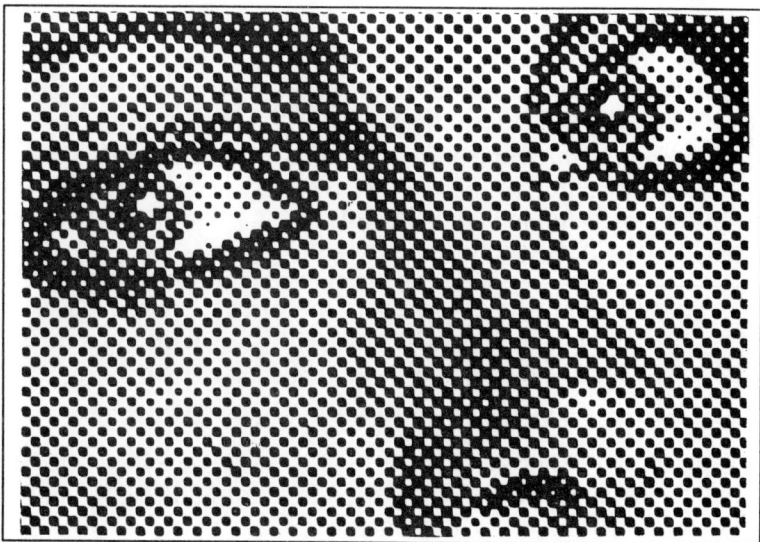

View both of these examples from a distance to see the tone effect.

Contact screen

There are three different routes to achieving a halftone conversion from a continuous tone image. They are:

Via a glass crossline screen.

Via a contact screen.

Via a computer and a laser beam.

We shall concentrate here on the contact screen because glass screens have no relevance to scanners, and are all but obsolete nowadays. The contact screen is used extensively in repro cameras and has seen use in scanners. Many scanners now use electronic dot generation (EDG) and the subject is covered later.

A contact screen is a piece of film upon which a dye image of a regular pattern of density differences has been created. It is called a contact screen because it is held in very close contact, usually by vacuum, with the photographic film on to which the image is recorded. The halftone contact screen works by modulating the light falling on to the photographic film in relation to the quantity of light reaching the screen. The continuous tone image, multiple shades of grey, is focused upon the photographic film, and passes through the contact screen.

Figure 2.7 shows a typical contact screen pattern. Whereas at first glance the pattern seems to be made of sharply defined light and dark dots, a closer inspection of the density differences would show a gradual change between the high and low density areas. This gradual decrease in density from the centre of a dark area out to the centre of a light area is known as the profile.

Fig. 2.7 Typical contact screen pattern

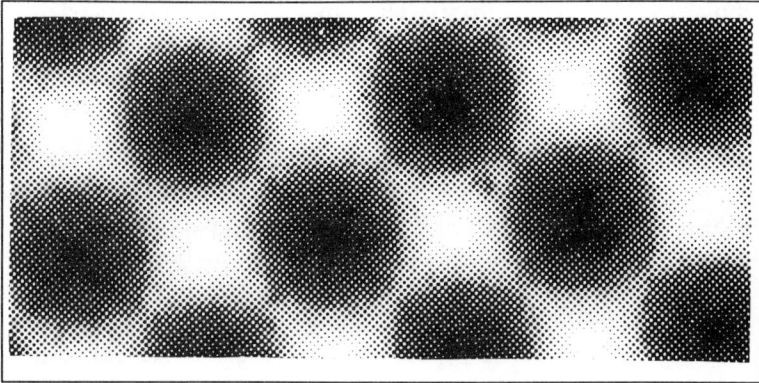

Figure 2.8 shows a highlight, mid tone and shadow representation taken through to a negative and how varying quantities of light, or brightnesses, are translated into dots of varying diameters.

Fig. 2.8 Tone representation

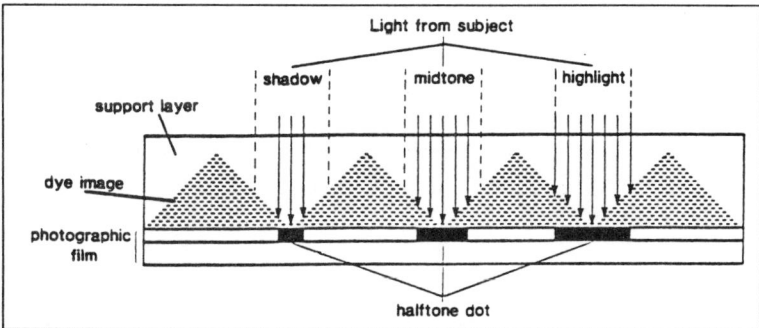

The screen must always be used with the dye image side in contact with the photographic film. Failure to do this will result in incorrectly formed dots. The dye image side can be recognised by the fact that it is matt, whereas the support layer side has a gloss finish.

Screen ruling

Simply stated, the finer the screen ruling, the more lines per centimetre, and the finer the detail which may be reproduced on a high quality paper surface. It will be a customer preference which screen ruling to use and this requirement will be based upon the printing process and the materials used.

Stemming from the days when screens were made as ruled lines on a glass plate, the screen ruling refers to the number of cells per linear centimetre of the screen. A 60 line screen ruling will produce 60 dot centres per linear centimetre of the image.

Typical screen rulings, in lines per centimetre, are given in Table 2.1.

Table 2.1

Typical screen rulings

Lithography	Letterpress	Gravure	Paper type
40	26	40	Newsprint
48	40	60	Machine finished
54	48	66	Process coated
70	60	79	High quality art

Correct notation is in the number of lines per linear centimetre of the screen, but it is equally likely to be referred to in lines per inch (see Table 2.2).

Table 2.2

Screen ruling equivalents	
Lines per centimetre	**Lines per inch**
20	50
26	65
30	75
40	100
48	120
60	150
70	175
79	185

Determining the ruling value of any particular ruling can be achieved using a Screen Ruling Tester. This is a readily available device which is made on a piece of film which is laid over the screen. By rotating the tester and noting the patterns generated, the screen ruling can be identified.

Screen angle
The screen angle refers to the angle formed by the direction of the dot structure in relation to the vertical finished position of the picture. Note that a line of the low density areas follows an angle of 45° to the horizontal side of the illustration. This screen would be described as a 45° screen.

Fig. 2.9 Screen angle

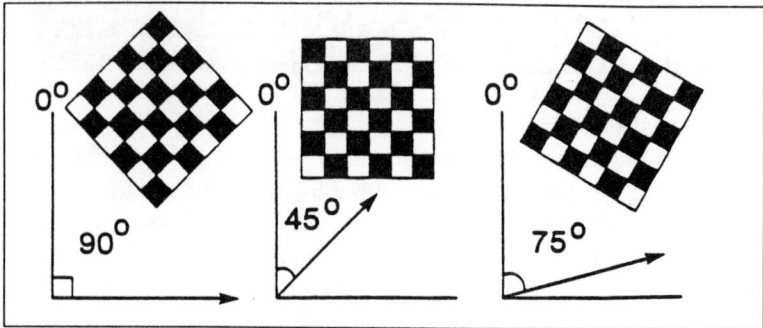

The screen angle quoted is usually measured from the horizontal. The angle at which the halftone pattern is least obvious to the eye is 45°. At this angle it becomes far less noticeable than when the rows of dots are aligned horizontally or vertically. So for any illustration which is to be printed only with one ink then a 45° screen will be used.

Screen range
A contact screen is only capable of reproducing a limited range of the greys of the original. The range that a particular screen can cover is determined at the time of its manufacture. It is the difference in density between the minimum and maximum areas of a screen dot, which determines the screen's reproducible density range. The screen range refers to the maximum density range of an original which that particular screen is capable of reproducing as a halftone.

Recognition of halftone values
The size of a halftone dot is expressed as a percentage. The reference percentage is to how much of a given area is covered by the dot. A 5% dot area means that 5% of the film or paper is covered leaving 95% open. A 50% dot area means that exactly half of the area is covered and a 75% dot will only leave 25% open.

Figure 2.10 shows enlarged versions of different dot percent areas so that the visual relationships can be compared.

Fig. 2.10 Halftone dot percent areas

Dot area assessment

A magnifier with an enlargement of about x10 is an essential tool
for checking dot percentages on film. However, if individual dots
need to be studied for shape and formation then a good quality
dot microscope with a minimum magnification of x30 should be
used. For an accurate measurement of dot area a densitometer is
necessary. Table 2.3 shows the relationship between measured
density and halftone dot area.

Table 2.3

Relationship of density to dot area

Density	Dot%	Density	Dot%	Density	Dot%
0.00	0	0.18	34	0.70	80
0.01	2	0.20	36	0.75	82
0.02	5	0.22	40	0.80	84
0.03	7	0.24	42	0.85	86
0.04	9	0.26	45	0.90	88
0.05	11	0.28	48	0.95	89
0.06	13	0.30	50	1.00	90
0.07	15	0.35	55	1.10	92
0.08	17	0.40	60	1.20	94
0.09	19	0.45	65	1.30	95
0.10	21	0.50	68	1.40	96
0.12	24	0.55	72	1.50	97
0.14	28	0.60	75	1.70	98
0.16	31	0.65	78	2.00	99

Applying halftone and colour principles

Now we understand the principle required to produce a colour reproduction we must discuss the method of applying the half-tone structure to the images and separating the original photograph into its three constituent separations.

Figure 2.11 shows the principle of separation. The original copy is exposed successively through three filters, to separate the red, green and blue content of the copy. All the colours transmitted by the filter cause the film to blacken after development, whereas the colours not transmitted obviously leave the film unaffected. Thus via the red filter all the colours containing red (white, red, yellow and magenta) give a developed image. Positives are then made of this and these are shown in the diagram as the printed plates (see also Plate 6). We could, of course, make photographic positives and reform the original additively by placing each in a projector with the same filter as that used for separation. However, as already discussed, additive methods cannot be used for reflection prints, and we need a subtractive system.

Fig. 2.11 The principle of separation

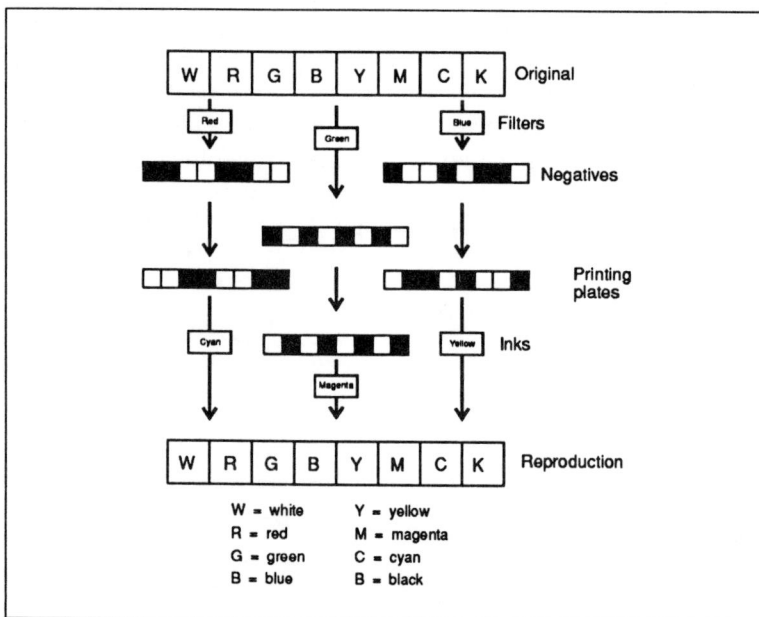

W = white   Y = yellow
R = red   M = magenta
G = green   C = cyan
B = blue   B = black

Each plate is a reverse image of the negatives and thus gives an image area which is the record of the copy which is not red, not green, or not blue. By printing these respectively in the inks which absorb red (cyan), absorb green (magenta) or absorb blue (yellow) we obtain the required result. At a first glance it is not always obvious why we cannot expose the original through cyan, magenta and yellow filters and print in red, green and blue, or alternatively produce printing plates with the image areas corresponding to the negatives and print these in red, green and blue respectively. The answer lies in the fact that inks subtract light from white paper and by setting up a diagram such as Figure 2.11 with these parameters it can be seen that such a procedure will not work. If the primary colours are red, green and blue, any attempt at overprinting two of these to obtain a secondary will result in black and not cyan, yellow or magenta.

Finally we need to consider colours other than the primaries or secondaries, and in Figure 2.12 we return to the example of a saturated orange. This is a colour which reflects red light strongly but green less. (If it reflected equal amounts of each it would be yellow). Thus if it is exposed through the three filters and a halftone screen the amount of light falling on the film through the red filter is sufficient to produce a 100% dot (solid) but since less green light is reflected, less is transmitted by the filter and only a 50% dot results. Since no blue light is reflected no exposure occurs. By making plates and printing we add a solid yellow ink to a 50% tint of magenta which absorbs all the blue light, half the green but no red. Thus the eye receives a mixture of red light and rather less green, resulting in orange.

Fig. 2.12 Reproduction of a saturated orange

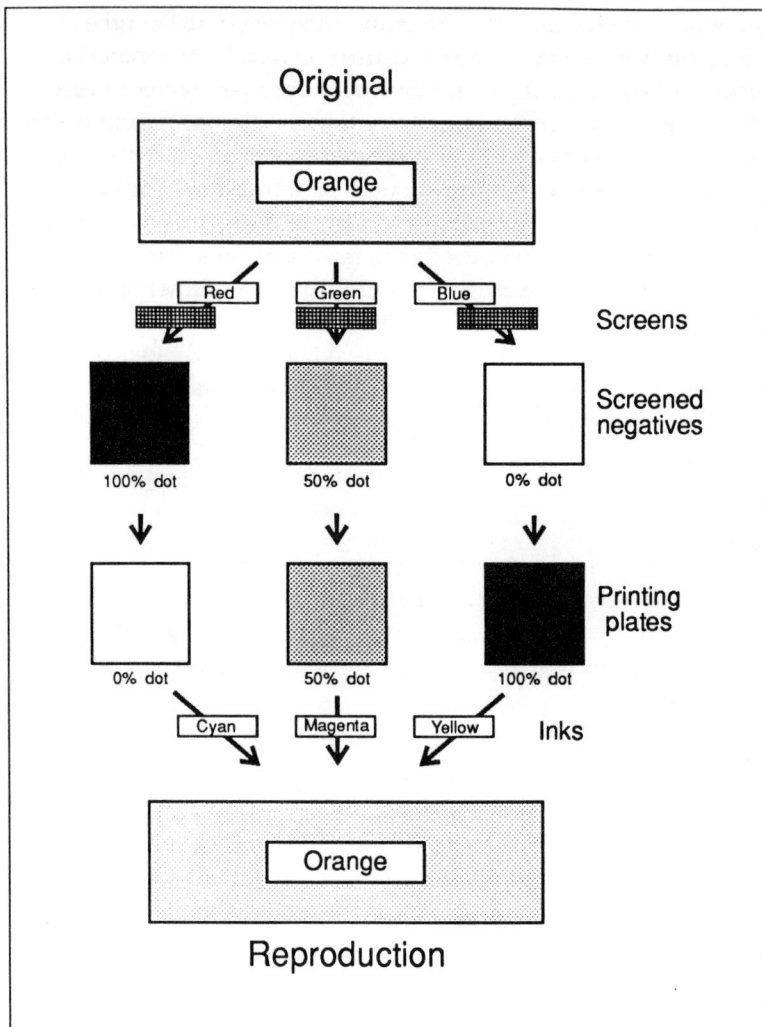

Principles of densitometry

The use of density as a unit of measurement follows directly from
work undertaken in the latter part of the 19th Century. It was

known that the photographic process was possible due to the reduction of silver halides to atomic silver in the photographic materials. When exposed to light and developed, the degree of reduction was related to the exposure received by the material. Because of the difficulty of measuring the mass of silver reduced directly, it quickly became apparent that a different system was required in order to make objective measurements. The obvious property to measure was the attenuation of light transmitted or reflected by the photographic material, since that is the effect in which we are interested, and in the early days this was the simple measurement which was carried out, using the following formulae:

$$\text{Transmission factor} = \frac{\text{Light transmitted by the exposed material}}{\text{Light transmitted by the unexposed (clear) material}}$$

$$\text{Reflection factor} = \frac{\text{Light reflected by the exposed (or printed) material}}{\text{Light reflected by the unexposed (or unprinted) material}}$$

However, in a classic paper of 1890, Hurter and Driffield pointed out that the mass of silver reduced was not directly proportional to the transmission factor, but should be proportional to the logarithm* of the opacity, and this was termed density. This is defined as:

$$\text{Opacity (O)} = \frac{1}{\text{Transmission factor (T)}} \quad or \quad \frac{1}{\text{Reflection factor (R)}}$$

$$\text{Density (D)} = \log_{10} 1/T \quad or \quad \log_{10} 1/R$$

*For those not familiar with logarithms they are described in an appendix.

The reason why density is proportional to the silver present may be understood from the following example.

Imagine light to be incident on three layers of photographic material, as shown in Figure 2.13, each layer of which only transmits one-tenth of the light falling upon it. Let us assume that we start with 1000 units of light falling upon the first layer. This layer will transmit 100 units of that light, since T = 1/10. Similarly the second layer will then only transmit 10 units, and the third 1. Thus, the total light transmitted is one-thousandth of that falling upon the first layer, and the total opacity is thus 1,000. This result is achieved by multiplying together the opacities of each layer separately (*ie* 10 x 10 x 10 = 1,000, which may be described as 10^3).

Fig. 2.13 Light transmission by photographic emulsion with T = 1/10 or opacity = 10

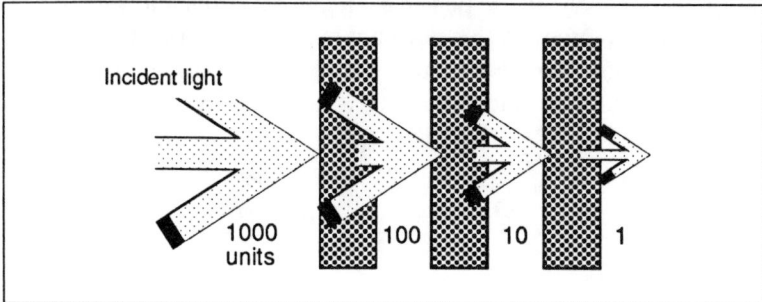

Mathematically it is convenient to express such light absorption logarithmically, since the logarithm of the opacity is proportional to the mass of silver present.

Thus, to summarise, when we talk of a density of 1.3, we mean that the light transmitted is $1/10^{1.3}$ times that incident upon the film. Thus, if 1000 units of light fall on a piece of film with this density then the light transmitted is approximately 50 units.

Certain important properties follow from the use of a density scale for measurement, one of the most important of which to us is the principle of additivity. In the example above we had three layers each with an opacity of 10 and hence a density of 1. When the three were combined we had a total opacity of 10^3 and hence density of 3. Thus, the total density is equal to the sum of the individual densities. Without such a scale, working out the total transmittance of a combination of three films, each with different opacities, would be more difficult, as may be seen from the following examples. Consider three pieces of film of density 0.2, 0.3 and 0.4. The total density is simply the sum of these, namely 0.9. The transmittances of these, however, are 0.63, 0.5 and 0.4

respectively and the total transmittance is given by the product of these, a relatively more difficult calculation. The use of logarithmic scales is also very convenient in the photographic industry for plotting the characteristic curve of materials.

It should by now be apparent why density as a scale of measurement has assumed such an important role in the photographic industry. The fact that it has a simple relationship to the amount of silver present, the principles of additivity and the convenient graphical analysis made possible, all combine to achieve this importance. However, this does not explain its application to printing, where we are measuring a pigmented ink in order to control the ink film thickness. This is taken up in Chapter 8.

Chapter 3 *The original*

Types of original

Originals may be in different physical forms, although basically of two types: a transparency, viewed by transmitted light; or reflection copy, viewed by reflected light.

Transparencies may be:

>*Original transparencies* – ie *the film that was in the camera to photograph the original scene.*
>
>*Duplicate transparencies* – ie *a copy of the above.*
>
>*Print film transparencies* – ie *made via a colour negative.*
>
>*Conversions* – ie *transparencies made from reflection copy.*
>
>*Colour negatives* – *but consult the trade house before considering their use.*

Reflection copy may be in the form of:

>*Photographic colour prints. (These must be made specially – ordinary 'en prints' will not do.)*
>
>*Dye transfers.*
>
>*Water colour paintings.*
>
>*Wash drawings.*
>
>*Poster colour paintings.*
>
>*Oil paintings, or other similar materials.*

Transparencies

A transparency is suitable for all methods of reproduction. The luminosity (or brightness) range of a reproduction will usually be less than that of the transparency and, therefore, some tone compression occurs in reproduction.

It is normal practice in graphic reproduction to keep most tone compression in the shadows, while maintaining good highlight contrast. If required, detail in particular parts of the tone scale can be retained specially or emphasised, but with consequent loss elsewhere. Clear, written instructions should be given in such cases. Advertisements, for example, call frequently for low key pictures in an attempt to convey some 'atmosphere', and it is important to communicate this concept to the trade house or printer.

The ideal transparency

The ideal transparency for reproduction has the following properties:

> Colour balance is neutral when viewed under standard lighting conditions (ISO 3664). Colour casts either overall or local are absent. Local colour casts tend to reproduce in a more obvious form and are difficult to correct. Overall casts on individual transparencies can be removed more easily, but still are best avoided since visual assessment is made more difficult.

> All the colours fall within the gamut of the combination of inks to be used. This can be judged from the printer's colour chart. Colours falling outside the gamut of the process will not be reproduced accurately. If this cannot be tolerated, extra colours must be printed.

> Freedom from dust, scratches, finger marks, stains, mottle and evidence of uneven processing.

> The image is sharp and has a fine grain structure, particularly if enlargement is envisaged. This will be dependent on the artistic effect required.

> It has been exposed to give a highlight density within the range 0.3-0.55. The highlight area referred to in this context is that containing the lowest density in which image detail is to appear in the reproduction. Catchlights will have a lower density than specified above. In order to retain detail in the reproduction of a high key subject, the transparency should have a highlight density in the

range 0.5-0.6, *ie* slightly higher than normal. A high key describes a subject in which the majority of important tones are in the highlight to mid-tone part of the tone scale. Over-exposed transparencies will lack colour saturation and have 'burnt out' highlights (*ie* containing no detail). Under-exposure gives poor shadow detail.

The shadow density is within the range 1.8-2.8. This is the density of the deepest shadow in which image detail is to appear in the reproduction. Frequently, original transparencies will have higher densities and these areas will reproduce as solid colour not containing detail. The maximum density at which it is required to reproduce detail should be stated, if possible.

It is the same size as the reproduction which is required. However, size changes are a normal part of the reproduction process. The printer's specification should give details.

It is wrong reading, *ie* the image is reversed left to right when looking at the emulsion side. All original transparencies will naturally meet this requirement. Duplicate transparencies may not.

Duplicate transparencies
Original transparencies may be duplicated for the following reasons:

To enlarge or reduce a transparency to the size required for the reproduction, or to enlarge a small portion of a transparency to final size. Image quality and subject matter can then be easily visually assessed.

To bring a number of original transparencies to the same end densities.

To remove colour casts.

To make several duplicates so that simultaneous publication in a number of places may occur, or to protect an irreplaceable original from the rigours of reproduction, *eg* photographs taken on the moon's surface.

To correct local colour casts by hand retouching and make general alterations. This is done more safely on a duplicate than the original.

To produce page ready copy.

Page ready copy
This may be known also as camera ready copy or scanner ready copy. It consists of an assembly of several transparencies planned to final layout and size. Each transparency is normally a duplicate since they must all have the same end densities, and the same colour balance. It is advisable to avoid mixing original transparencies with duplicates since they are likely to have different optical properties.

The emulsion of the individual elements should always be on the same side. Mounting transparencies together is a very skilled operation and is best left to a specialist.

For more complex assemblies, where, for example, one image is to be inserted into another, the following techniques are available:

Cutting and butting. A relatively simple technique, but should be used only where the subject to be inserted has well defined edges of simple shape.

Emulsion stripping. This is more difficult but can cope with finer detail. The cut edge is not so likely to be apparent as with cut and butt techniques.

Photographic montage. This is the most expensive technique and is used usually when much retouching would be necessary otherwise.

Some colour scanners can also do certain types of *combination work*.

Before any duplication or montaging of transparencies is undertaken, consult the trade house or printer to save unnecessary work.

Reflection copy

Recommendations for the production of reflection copy depend somewhat on the facilities available at the trade house or printer in question, and reference should be made to their specifications.

When there is a choice it is recommended that reflection copy be produced 1.5 to 2.0 times final size, unless the original is for small reproductions

Photographic colour prints

Prints can be made either from negatives or positives but which-ever system is used they should be made from the original negative or positive of the subject. Prints should not be copies from another print and they should be made specifically for the purpose of reproduction. When this is not the case it often puts them into the category of problem originals.

Colour prints for amateur consumption are made as cheaply as possible because of the intense competition in the industry. As such the final quality is likely to be the minimum the customer will tolerate rather than the maximum the process is capable of. Due to unavoidable losses inherent in the printing process, it is only possible to achieve high quality print economically from high quality originals. To qualify as appropriate, photographic prints should, at least, have the following properties:

Be clean and damage free.

Gloss surface material.

Be correctly exposed.

Be sharp.

Within subject requirements, be neutrally colour balanced.

Prints with regular patterned surfaces, such as silk or canvas finishes, can cause problems due to the surface finish conflicting with the halftone screen pattern. This is one reason why a gloss finish is preferable. Another reason is that fine details can become obscured by a coarse surface pattern.

Because of the nature of a print and the controls available during its manufacture, the scanner operator should be able to presume that facsimile reproduction is required.

Dye transfer prints

The dye transfer process is a photographic technique which produces high quality, continuous tone colour prints, but follows a sequence of separation very similar to process printing.

Dye transfer prints are much more costly to produce than conventional colour prints, but because of the multiple stages involved in their manufacture, they do have advantages when considerable retouching is needed. Tone range alteration, colour retouching and all forms of hand work can be undertaken readily on each of the separations. These are facilities which the graphic reproduction industry takes for granted but are not usually available during the conventional production of a continuous tone colour print.

A dye transfer print is likely to be the highest quality original ever supplied for reproduction. As such it should present the least problems.

Water colour paintings

A water colour painting can be reproduced satisfactorily but is most successful if the substrate matches that of the original. Conversion to transparency form should be avoided since differentiation between highlights and light tints will be lost.

Wash drawings

Where these have an outline drawing, this should, if possible, be drawn on a separate overlay rather than on the same surface as the colours. The line drawing can then be printed using a single colour which avoids registration problems. All hand drawn originals should be produced on ordinary white board or paper, not on fluorescent white surfaces. At the colour separation stage, the latter reproduce pink, which is not discovered sometimes until proof stage.

Mixtures of poster colours, oils and water colours on any one original should be avoided and fluorescent paints and dyes not used.

Oil and poster colour paintings

The main difficulty with paintings of this type is the long density range and existence of non-reproducible colours. It is common practice to use more than the four process colours and this should be discussed with the printer.

To ensure that the reproduction of an oil painting has the required surface texture effect, special attention should be given to lighting angles when the colour separations are being made, or when a transparency is taken of the painting. Unwanted surface reflections which desaturate colours and cause loss of shadow detail should be avoided.

Conversions

Reflection copy is sometimes converted to transparency form as the first step in the reproduction process. The reasons are similar to those necessitating duplicate transparencies.

Mixed originals

Occasions arise when different sorts of originals are to be integrated into, say, a brochure. The end densities and spectral qualities of the dyes, pigments and substrates in these originals are unlikely to be the same, and satisfactory duplication to one common form may not be possible. The trade house or printer should be consulted before proceeding.

Retouching

Whenever originals are retouched there is the possibility of metamerism. If metamerism does occur the retouched areas will appear correct on the original but any subsequent reproduction will show errors.

In the case of photographic originals, metamerism can be avoided by using only those dyes recommended by the manufacturer of the material concerned. In particular, colour prints should not be retouched with poster colours since metameric effects are almost inevitable.

Principal factors to be considered when installing standard viewing conditions for the visual assessment of colour

The following summary briefly outlines the principles to be considered when installing standard lighting. More detailed descriptions are given in an appendix and specified where appropriate. It must be stressed that this is only a guide and specialist lighting engineers should be consulted before any installation is undertaken.

Ensure that the chromaticity and spectral distribution of the illuminant conforms to that laid down in the appropriate standard, *viz* BS 950 (1) for colour matching and appraisal and BS 950 (2) or ISO 3664 for viewing in the graphic arts.

Ensure that levels of illumination and luminance conform to the standards.

Ensure that no extraneous light is permitted to degrade the specified spectral distributions.

Ensure that no glare, either directly from the light sources or by reflection from glossy surfaces, interferes with the field of vision.

Ensure that all immediate surround areas are mid-grey in colour.

Ensure that no observers with defective colour vision make colour assessments.

Ensure that the observer is not visually adapted to a different mode of viewing.

For assessments made under BS 950 (1), ensure that metameric effects may be assessed.

Photographic origination for graphic reproduction

Creation of the original

In any normal commercial environment, choice of subject matter is rarely, if ever, left to the discretion of photographers. They are given, usually in very specific terms, exact instructions as to what is required. The photograph will have a purpose and it will be a measure of the photographer's skills and the instigator's descriptive powers as to how close to the initial requirements the final photograph remains.

Having been set a particular task, be it illustrate a news report, advertise a product or entertain the public, the treatment of that task is often under the control of the photographer. Choice of camera format and film size, type and make of film stock, choice of lighting, degree of exposure, often angle of view, choice of perspective and framing within the format used are all areas in which the photographer can influence the final outcome whilst still working within the tight constraints of an art director's brief.

Of the areas listed, where the photographer does have the opportunity to affect the final outcome, the ones which concern the quality and success of the graphic reproduction most directly are:

Choice of film size.

Type of film stock.

Choice of lighting.

Degree of exposure.

Other points are concerned with the aesthetics of the picture and, as such, are not dealt with here.

Before leaving this first topic of 'subject matter', there is one aspect of accurate reproduction that is often overlooked in the heat of the photographic studio. That is, how can the subject itself, be it printed box, colourful garment or even natural bloom, conspire with the film chosen to defeat the photographer's attempts to reproduce what they see? There are a number of circumstances which can cause a film not to reproduce colours as seen by the human eye. Like the human eye, colour film responds to red, green and blue light reflected from the subject. However, film cannot match the flexibility and complexity of human vision. In order that, under the greatest number of 'normal' conditions, a film's rendition of colours is considered correct, certain assumptions are made by the manufacturer. Basically, these are that flesh tones are of prime importance and that neutrals (whites, greys and blacks) should remain neutral. Other important common colours are blue sky, green grass and yellow sand. Optimising the film's handling of these colours imposes limitations in other areas. For example, some shades of lime, pink and orange might not reproduce exactly. It would be possible to produce a film that could improve on these colours but then the more important ones would suffer. For most subjects it is sufficient that the film approximates the same wavelength-sensitivity relationship as the eye. All that is required is for the accumulated effects of the red, green and blue light to be in the same ratio in the film as they are to the eye. However, some of the most pronounced reproduction differences are caused by the fact that photographic emulsions are sensitive to wavelengths outside the visible spectrum. These difficulties arise when photographing subjects which exhibit ultra-violet reflectance, ultra-violet fluorescence or infra-red reflectance effects.

We shall take these three factors separately. Firstly, ultra-violet reflectance. Any subject which reflects ultra-violet radiation efficiently will reproduce bluer in a photograph than it looks. This is because photographic emulsions are sensitive to ultra-violet but the human eye is not. If the subject is blue to begin with then the effect is likely to go unnoticed but not with subjects of other colours. Additional blueness may distort the rendition sufficiently to make the photograph unacceptable. Blueness caused by reflected ultra-violet radiation can be overcome by a filter over the lens which will prevent reflected ultra-violet from reaching the film. If the subject is studio lit then, by filtering the light sources and preventing ultra-violet from reaching the subject, the same end can be achieved.

Ultra-violet fluorescence effects are caused by parts of the subject absorbing ultra-violet radiation and re-emitting it as visible light.

Many white fabrics and papers contain optical brighteners and there are many coloured papers available with fluorescent coatings. As the eye is not very sensitive to this part of the spectrum, the problem may not be obvious until after the photograph has been taken. If it is suspected that an unwanted blue shift in the picture has been caused by fluorescence, examining the subject under a ultra-violet source will make it obvious. Fluorescing areas will stand out clearly. This cannot be cured by a filter over the lens, as the radiation has already been converted to visible light by the subject. The light sources must be filtered to exclude ultra-violet radiation.

Finally, infra-red reflectance effects. The human eye has virtually no sensitivity in this area of the spectrum but colour films do. The classic example of this problem is the blue 'morning glory' flower which, under most lighting conditions, photographs as pink. Some types of organic dyes used by fabric manufacturers, particularly on synthetic materials, have this high red reflectance. The effect of this reflectance is most noticeable in medium to dark green fabrics. The photographic effect is to neutralise the green and, in extreme cases, the reproduction would appear a dark muddy brown. By examining the subject through a deep red filter under tungsten lighting, an evaluation can be made of the likely photographic results. A green, natural fibre material will look black whereas a fabric with high red reflectance will appear lighter. A side by side comparison of the two will forewarn of a likely reproduction problem. The reproduction can be improved with alterations to the light sources or by the use of complicated filter packs, but more often than not it is left to the skills of the graphic repro retoucher/masker to save the day. It is the photographer's responsibility to their client, however, to be aware of where the problems are arising in their section of the reproduction chain and to minimise the amount of after work needed.

Choice of film size
When considering the optimum size of a transparency destined for reproduction, in general 'biggest is best'. This cursory statement leaves many questions unanswered. Imagine the photojournalist's reaction to the suggestion that they return from an assignment at Heathrow airport with a nice selection of 10x8in colour transparencies. They would be superb quality but, without

the backing of a Hollywood size film crew, they would have missed the 'moment' which the editor may require for the feature. The reverse is, of course, true. No international fashion magazine is likely to run as its front cover a shot taken from a 110 size negative. A more realistic answer would be, within the constraints of economy, location, subject availability, time and general practicability, then biggest is best.

There are a number of reasons why the larger the original the better the reproduction is likely to be. Firstly sharpness. If a 35mm is used then there is a very good chance that the reproduction will be an enlargement of only a portion of the original. This means that the slightest degree of unsharpness in the original will be magnified substantially. Couple this with the grain structure which is present in all photographs and there are real problems. Colour film, due to the fact that it is a sandwich of three emulsions, is inherently less sharp than its thinner colleagues in black and white. Grain and sharpness are two major reasons why the larger the original the better chance one has with the reproduction. However, there are other reasons too. Dust is one. Dust and damage will be most obvious in any large area of neutral tone. This is particularly well known by anyone working with 35mm transparencies in the audio visual field. From the moment the virgin film is placed in the camera it becomes prey for dust, sand, finger grease and all manner of foreign substances. No matter how careful the photographer is with the original it has to pass through so many hands before it appears finally in print that the chances of it remaining undamaged and clean are very slim. Accepting this as a fact of life, and because dust and damage do not increase in size in proportion to the original, the less enlargement needed for the reproduction then the less noticeable this sort of non-photographic defect will be. In the fortunate event of the reproduction being a reduction of the original then any problems will be reduced also. Other aspects of physical size are handling, retouching and contact masking. All these post origination treatments are simpler the larger the original.

Type of film stock – monochrome or colour, negative or transparency?
It is very unusual for black and white originals intended for reproduction to be taken on reversal film. Even though this type

of film has particularly fine grain characteristics and excellent tone reproduction qualities, the advantages of using a negative to positive process easily outweigh these good points. In terms of initial exposure latitude, contrast control both of the negative and the positive, opportunities for shading, dodging and retouching, all give the photographer an enormous amount of flexibility over the finished result. Points to remember when choosing an original negative stock for monochrome work are:

> *The slower the emulsion speed the smaller the grain size and, therefore, the less noticeable the image graininess on enlargement.*

> *The slower the film speed the thinner the emulsion and the sharper the image will appear.*

> *The slower the film speed the higher the contrast.*

In colour, the greatest proportion of photographs taken for reproduction are supplied to the printer as transparencies. As with monochrome, colour prints are made to size easily, re-touched, montaged and assembled into page layouts, but prints often have poorer colour saturation and definition than transparencies. Colour negatives may be used for reproduction but they have problems of their own. Colour negatives, even for experienced photographers, are difficult to assess visually. The in-built colour mask gives them an all over orange appearance. Trying to assess negative colours as well as negative tones, makes them useless as a visual guide to the end result. The photographer will need something to prove that the work is correct. The client will want to know how the picture will look. These are good reasons for having a final and correct transparency to pass to the printer. There are many suppliers of film offering a range of different colour reversal films on the market today. Some of these films are formulated for very specific tasks such as photomicroscopy, infra-red sensitivity and duplicating. Films which satisfy repro needs can be found providing that the needs are known, and it is understood how a particular film needs to be handled in order that it can fulfil its purpose.

Choice of lighting

There is the choice of four major sources of light – natural, tungsten, flash and fluorescent. Of the four, the most changeable, unpredictable and difficult to control for colour photography is natural light. It can vary tremendously in colour temperature and intensity depending upon the time of day, the time of year, the global location and the weather. Average colour temperatures are 5,500K for clear daylight, 7,000K for cloudy sky light, up to 10,000K for bright blue sky and shaded areas. For these reasons it cannot be assumed that as the film had 'daylight' written on the box and the photograph was taken out of doors then the colour balance will be perfect. Any photographer who takes transparencies consistently in natural light for reproduction should own a colour temperature meter and a set of 'mired' filters, as well as an exposure meter.

A convenient source of artificial illumination is tungsten lighting. This can be split into three main types: domestic, as would be found in the home and in a few office and industrial environments; overrun tungsten, which is specifically for photographic use; and tungsten halogen, again for photographic and video studios, shop display and some security surveillance situations. Domestic tungsten need not really concern us here other than to be aware of its relatively low colour temperature, between 2,600- and 2,800K, and to watch out for its presence in mixed lighting situations. Overrun tungsten bulbs, such as Photofloods, are designed to have high output and a colour temperature of 3,400K. Unfortunately, because of the nature of their construction, they have a very short working life.

For critical colour work, a No. 1 Photoflood will have an effective working life of only 1 hour. A No. 2 will last for about three to four hours. For this reason, and also because of their low initial cost, they are more often used by amateurs than professionals. Tungsten halogen lamps do not suffer from this short working life, providing they are correctly handled, and their colour temperature is 3,200K. This is the correct level for artificial light balanced film. For critical colour work some form of voltage stabilisation should be used in order to guarantee correct colour output. A wide variety of lamp sizes and styles are available.

These range from tiny, intense focusing spotlights, for placing highlights and rim illumination in specific areas, through to enormous diffuse arrays which give the effect of natural northlight illumination (but not northlight colour) with very soft shadows. Some disadvantages of incandescent lighting such as this is that it generates a great deal of heat, and can make a studio a very uncomfortable place to spend any time. Tungsten halogen lamps are fairly expensive to replace and whilst hot are very susceptible to movement damage. Because the temperature of the lamp envelopes gets so high, any finger grease on the lamp can set up intolerable strains and the lamp may blow.

Flash illumination can be divided into bulb or electronic, hand-held or studio. In terms of colour temperature, all flash is balanced for daylight type film between 5,000- and 5,500K. Flash bulbs are waning in popularity because of their high cost and recent developments in electronic flash, but there are still some aspects of professional photography which require their use.

Both expendable and electronic flash have advantages over tungsten illumination. For example, colour temperature is consistent with flash regardless of supply voltage and flash is virtually cold during operation.

In the area of hand-held use, the flash bulb has given way entirely to the electronic flash gun. Speed of use combined with automatic exposure determination, the increase in quality and speed of emulsions allowing the use of less powerful flashes, the flexibility of modern hand-held flash heads with adjustable output angles, and the ability to tilt the head for bounced flash, have helped their popularity. In studios, electronic systems are highly sophisticated with integral power supplies, an impressive array of reflectors and baffles, and with the use of photoelectric couplers the need for masses of trailing wires has largely been eliminated. Flash is ideal for photographing food which can be affected easily by heat, for fashion and portraiture, and in still life studios, where colour consistency is of paramount importance.

Finally, fluorescent light, as a source of light for photographers, can be excellent for black and white and a real headache for colour. Locations such as factories, offices, supermarkets, *etc*, are lit invariably by fluorescent tubes. The effect is an even, diffuse lighting which, although it is all coming from above, and, therefore, top surfaces will be far brighter than sides or under surfaces, can be used successfully. However, due to the nature of the spectral emission from these sources, their effect on colour film is to produce a distinct green cast. There are manufacturers offering correction filters for use in these lighting conditions, but no one filter is correct for all fluorescent situations. Kodak publishes 18 different filter combinations for the most commonly used tubes with its ranges of negative and transparency films. The professionals who expect to encounter fluorescent lighting will carry a range of ready made filter packs and will do a test run to ascertain the perfect correction combination. If this is impracticable, they will at least find from the maintenance crew of the location what type and make of tube is in use, then choose the correction pack accordingly. Note that many locations, such as supermarkets and department stores, are likely to use different types of fluorescent lamps in different parts of the store.

Any location photographer will regularly meet mixed lighting situations, *ie* a subject lit with two or more of the lighting types mentioned. For black and white this is no problem and for colour, daylight and flash will mix. Any other combination will cause colour inconsistencies. The photographer has three choices:

> *Remove all but one source.*

> *Balance the sources with filters so that they match.*

> *Use negative stock and attempt to correct the differences in printing.*

This third should be tried only as a last resort if the other two solutions are impossible. For the reasons mentioned earlier, transparencies are preferred and this solution, if performed properly, will add enormously to the cost.

Before leaving lighting we should discuss style and lighting arrangement. Here we return to the topic of aesthetics and customer requirements. The contrast, mood and texture of any shot, studio or location, is set by the choice and handling of the lighting arrangement. We could discuss at great length the 'orchestration' of the overall image combined with the 'subtle invocation of minor emotions', but really that is not our concern. There are very simple and straightforward rules which apply to lighting for graphic reproduction. A transparency is capable of achieving a density range in the order of 3 whereas a litho reproduction printed on coated paper would have a typical density range of only 1.8. The range must be compressed and normally the printer will go for a separation which will hold the highlight detail. Therefore, it is the shadow end of the range which will be most compressed. Important detail should be kept out of areas of shadow in the photograph. For best results on paper keep the lighting flat. The only areas of clear film should be catchlights (sparkle points). Shadows should be light and diffuse. Fill-ins, white reflectors, umbrellas and large area lights are all allies in the fight against contrast. If contrast is needed to achieve an effect, then do not expect both ends of the tonal range to appear in full detail on the final print. With good feedback between the client, photographer and printer, it becomes a lot easier for the repro house to know which portion of the range they may compress and still retain the desired result.

Degree of exposure
For accurate reproduction, correct original camera exposure is critical. The two aspects which are most affected by incorrect exposure are end of the range detail and colour saturation. An over-exposed original, be it negative or transparency, will suffer from burnt out highlight detail. Skin tones will be bleached and anything above mid-tone may well be nearly solid on a negative or virtually clear film on a transparency. Negatives have a greater latitude to cope with over-exposure but transparency film should be within one-third of a stop of the correct exposure. Negative working also has the advantage of a second stage where corrections can be made. Also over-exposure causes desaturation of the colours in a transparency.

Under-exposure compresses the shadow end of the density range. Highlight detail will be retained but with gross under-exposure this may be all there is. Shadows will fill in and go solid on the reproduction. Because of the irreversible problems caused by over-exposing a colour transparency, of the two defects it would be preferable for the original to be a half to one stop under-exposed rather than an equivalent amount over-exposed.

Chapter 4 *Colour reproduction in practice*

In order to achieve good colour reproduction in the printing processes, we need to consider the application of the principles that were examined in previous chapters. We need to understand how the various principles affect each other, which parts of which principles are important, and perhaps to develop a new principle containing just the factors that apply.

Spectral considerations of a reproduction system
We have described in Chapter 2 the principles of a colour reproduction system using cyan, magenta and yellow inks and red, green and blue filters to separate the appropriate images to be produced by these inks. However, we have used the words red, green, blue, yellow, magenta and cyan very loosely.

The term red covers a whole multitude of colours and so we have to consider whether we can use any red, green or blue filter or any cyan, magenta or yellow ink, or whether there is some restriction, which means we have to use particular colours.

This is a very complex question and not very easy to answer specifically. In fact it is still an area needing research. However, in this discussion we shall examine some of the factors to be considered in the choice of inks and filters, and will start with inks.

Inks for colour reproduction

We shall start by ignoring any economic restrictions on ink purchase and consider what it is we are looking for in an ideal ink set. We know that they must be yellow, magenta and cyan and therefore absorb blue, green and red light respectively. We also know that this set of inks can, at least mathematically, produce any colour if we could devise some method of printing a negative amount of ink. This is obviously impossible, although by the addition of extra inks to a process set the result can be achieved.

However, for the time being we shall restrict our considerations to three colours only.

Thus our first consideration must be to choose an ink set which will reproduce as many colours as possible, or in other words produce the largest possible colour gamut. We have a method at hand (see Chapter 2) for measuring the gamut of colours produced by the addition of coloured lights and whilst not strictly applicable to subtractive colours, it does indicate the requirements. It utilises a colour diagram which is discussed in some detail in the Pira report *Colour matching and control* and can be considered a map of colours for our purposes. This is shown in Figure 4.1.

Fig. 4.1 CIE colour diagram

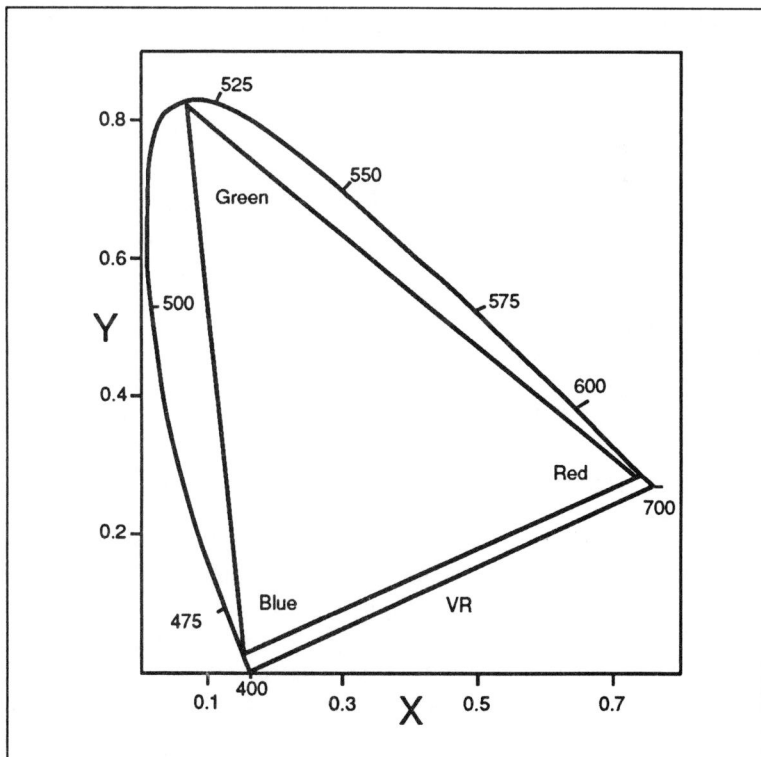

The diagram consists of a curved line which defines all the spectral colours (known as the spectral locus) and also a straight line VR (known as the alychne) which represents colours of red-purple, such as magenta, which do not exist in the spectrum but are obtained by mixing red and blue light. The compound line consisting of locus and alychne is a boundary for all real, 'seeable' colours; any other colour lies outside this.

Now, if we choose three points, representing coloured light sources, within this diagram and form a triangle joining them, we know that any colour within that triangle can be produced by mixing the three lights defined by the points. Colours between the triangle and the spectral locus and alychne can only be formed by the addition of a negative amount of colour. Such a triangle is shown in Figure 4.1, given by the mixture of spectral colours of approximately 650-, 520- and 450nm. The spectral distribution of these lights is shown in Figure 4.2.

Fig. 4.2 Spectra of light sources

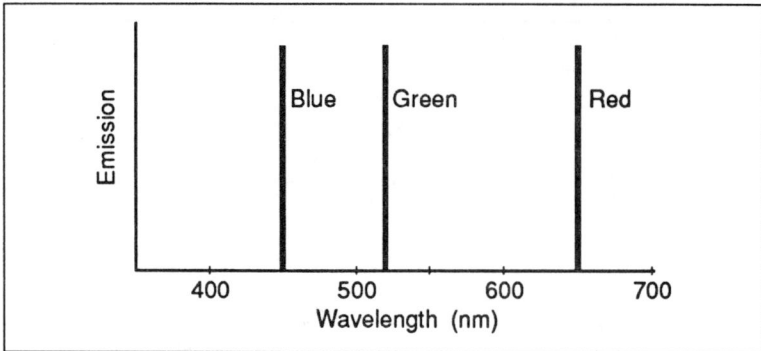

However, we can produce colour mixtures with red, green, and blue lights which are not narrow spectral lines as shown in Figure 4.2 but wider bands as shown in Figure 4.3. These are, in fact, the spectral distributions of a set of phosphors used in colour television.

Fig. 4.3 Spectral distribution of colour television phosphors

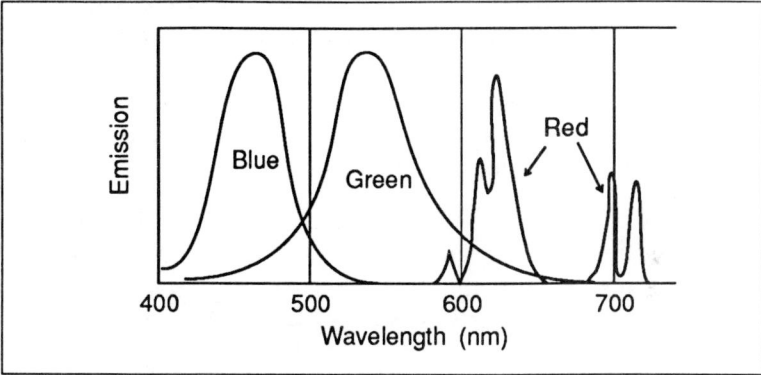

The point at which these fall on the diagram can be determined and the triangle drawn as in Figure 4.4.

Fig. 4.4 Colour television phosphors plotted on CIE diagram

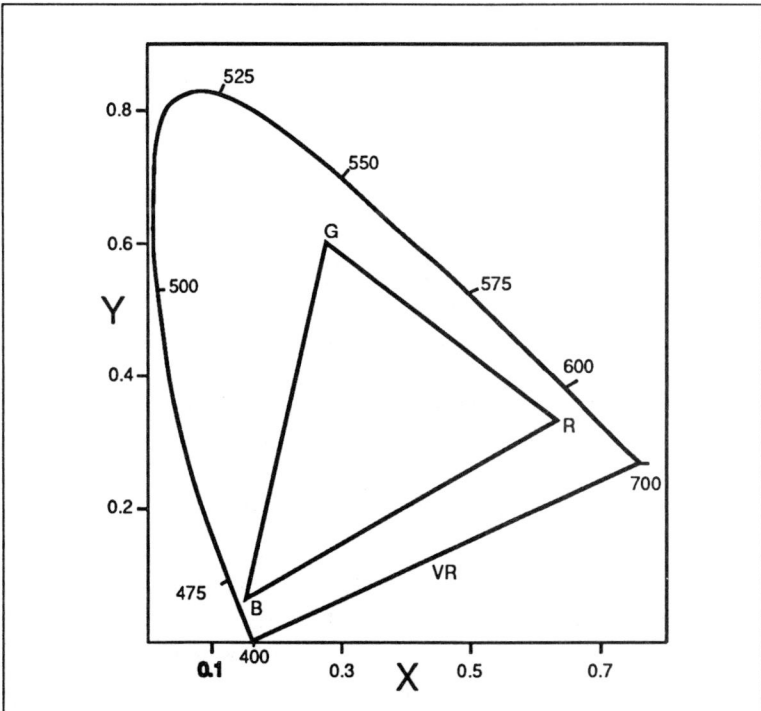

Immediately we can see that the colour gamut has been reduced. It could only be enlarged again by the addition of extra phosphors, (assuming that the three originally chosen are irreplaceable). Now we can apply the same sort of logic to our subtractive colours, and in Figure 4.5 we show the reflectance of a cyan, yellow and magenta ink set which can be considered ideal.

Fig. 4.5 Reflectances of ideal inks

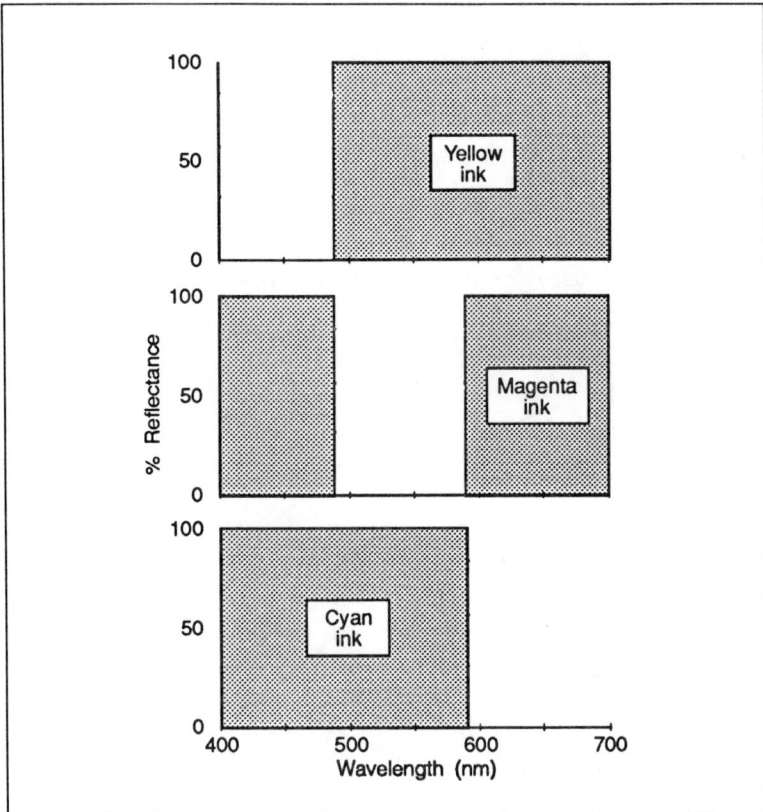

Each absorbs in one third of the spectrum and completely reflects in the other two thirds. If the absorption bands are narrowed, so that their boundaries do not touch, then at some wavelengths, no light will be absorbed and thus it is impossible to produce black. If on the other hand the absorption bands are wider, we have a situation where too much light is absorbed and all the colours appear darker than required. The actual wavelength intervals over which the absorption bands should occur is rather imprecise, but values of around 490- and 580nm are generally considered acceptable. By varying these, the actual gamut triangle will vary and the choice is to some extent dependent upon which colours are considered important in the reproduction. Our colour theory tells us that the three dyes are controlling red, green and blue blocks of wavelengths defined by the absorption of the inks. It is possible to plot these on the colour diagram and hence define the colour gamut they will produce. This is shown for the ideal inks in Figure 4.6.

Fig. 4.6 CIE diagram for ideal inks

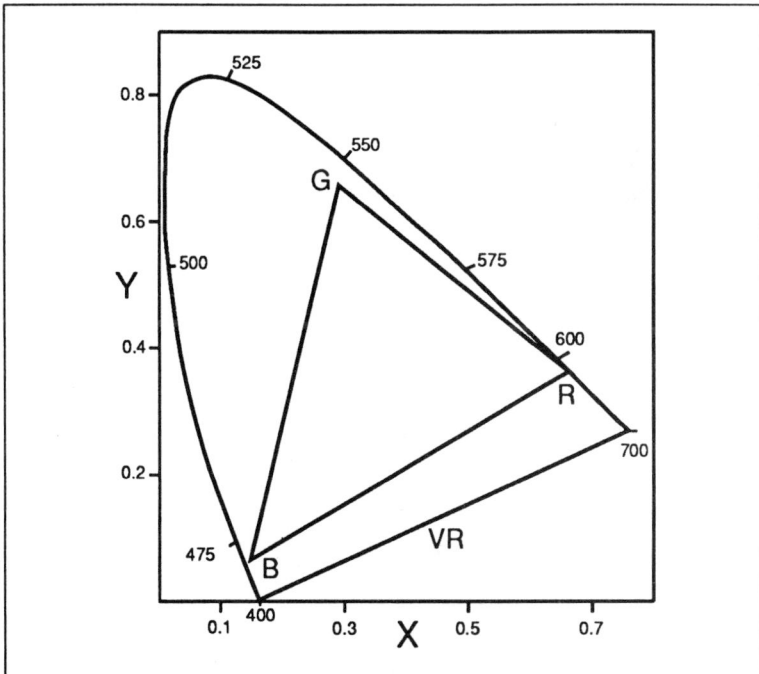

We can see here that many colours lie outside the reproducible gamut and this is a major limitation of the subtractive processes. We should bear in mind, however, that the occurrence of extreme

71

colours in nature is limited. The majority of natural colours to be reproduced are grouped towards the centre of the diagram. It is important that the reproducible gamut covers the majority of important recognisable colours, which should include skin colours, sky colours, foliage colours and the like. As stated earlier, variation of the boundaries of the absorption bands will alter the points defining the gamut triangle and thus introduce colours into the gamut not included in Figure 4.6. However, this will invariably be at the expense of another area of the colour diagram. Thus the choice of primary colours must be, to some extent, a compromise, but obviously should be made so that the gamut is as large as possible.

Turning now to real inks we obtain spectral reflectance curves of the type shown in Figure 4.7. The four curves in each of the graphs of Figure 4.7 show the reflectances of different amounts of coloured ink.

Fig. 4.7 Spectral reflectances of real inks

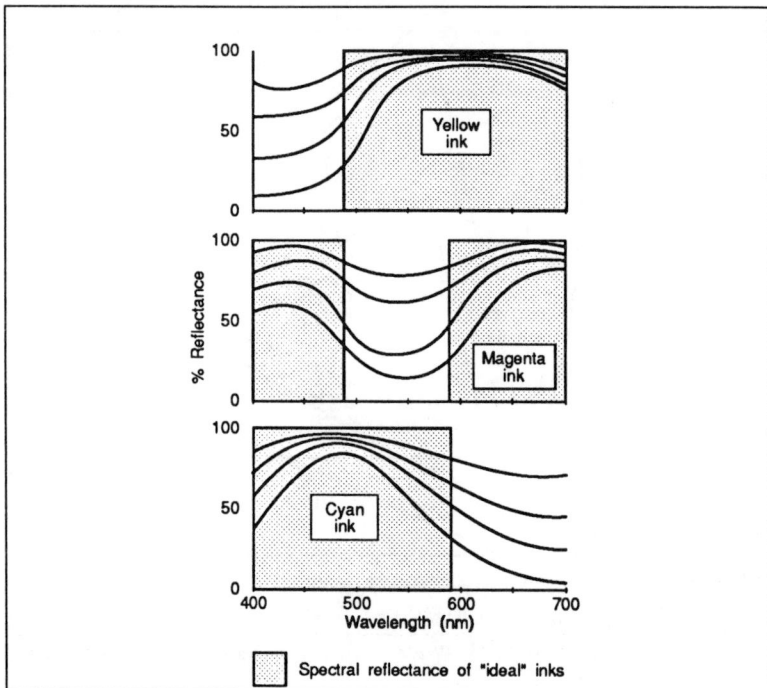

Spectral reflectance of "ideal" inks

As the ink film thickness is varied (as in the case of conventional gravure) or the halftone screen introduced (for other processes) two things are immediately apparent:

> *The inks have unwanted absorptions,* ie *relative to the ideal inks, they absorb in parts of the spectrum where they should be reflecting.*
>
> *The absorption bands have sloping sides.*

The effect of the unwanted absorptions is to darken the colours unnecessarily. This can be partly compensated for by colour correction and this will be covered later. The sloping sides, on the other hand, introduce an interesting effect. It can be seen that as the tone value increases, the absorption bands become effectively narrower. In theory this enables colours to be reproduced which lie outside the reproduction gamut produced by the ideal inks and this is shown by the broken line in Figure 4.8.

Fig. 4.8 The increased gamut of colours due to the sloping sides of the absorption bands

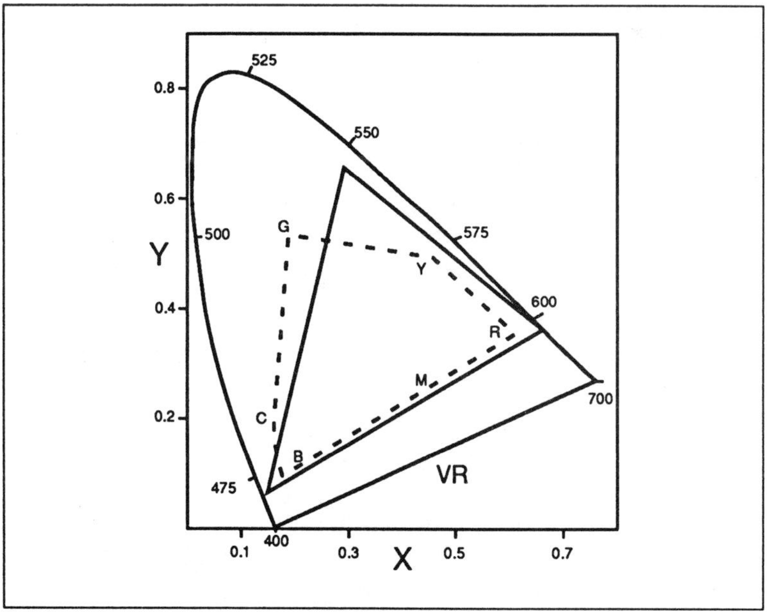

The full line is the boundary defined by the ideal inks, whereas the broken line is the gamut defined by the BS 4666 standard ink set. This has been obtained by plotting the values given in the standard for the primary and secondary colours and joining them with a straight line. In fact the lines will not be straight. If anything they curve outwards, thereby enlarging the gamut slightly, but this is adequate for our purposes. Calculation of the true gamut boundary between the points is somewhat tedious. We can see that the British Standard set produces a better gamut in the cyan-green region of the colour diagram than the ideal inks, but at the expense of the reds, yellow-greens and blues. A different set could be chosen which would improve one or more of these areas at the expense of another (or others). Thus in a sense real inks are better than the 'ideal' inks. However, there is one aspect of the colour gamut we have not yet considered, and that is the darkness of the colours.

Colour space is a three dimensional gamut, and the diagram on which we have shown gamuts thus far only shows two of these dimensions. The one excluded is lightness (or darkness depending upon which way we may look at it). Lightness would plot on an axis at 90° to the page surface. We have already stated that the darkness of the colours is increased by the unwanted absorptions and to show the complete picture we need to consider darkness also. As the darkness of a colour increases its apparent saturation decreases, and since the darkness of the BS 4666 colours is greater than the ideal inks, Figure 4.8 gives a somewhat misleading picture. Unfortunately it is not easy to show a three dimensional space on paper and so we usually only plot two dimensions at a time.

Several techniques have been devised to express colour in terms of three dimensions. The most successful of these, the Munsell system, and those devised by the Commission Internationale de l'Éclairage (CIELAB and CIELUV), result in colour spaces of spherical or cylindrical form. If we imagine a sphere having white at the top, black at the bottom, and the various colours (hues) arranged around the 'equator', then it follows that lighter colours occupy the upper part of the sphere whilst darker colours occupy the lower part. Greys fall naturally on the vertical axis, between white and black (see also Plate 3).

The mathematics for such spaces are necessarily complex and are not covered here. The order of colours around the equator is usually arranged so that complimentary colours are diametrically opposite each other.

Vertical slices through the colour space can be seen to expose to view a range of two colours, and white and black. Figure 4.9 shows a series of slices having white at the top and black at the bottom. The exposed colours of each slice are marked. (W* is a correlate of lightness.)

Notice that the sphere is somewhat distorted. In particular the lower diagram shows the yellow/blue cross section. Here, the yellow is high (light) and the blue is low (dark) on the lightness scale. The diagram shows that yellow is a light colour and blue is a dark colour – which is true.

Fig. 4.9 Two dimensional representation of slices through a three dimensional colour space

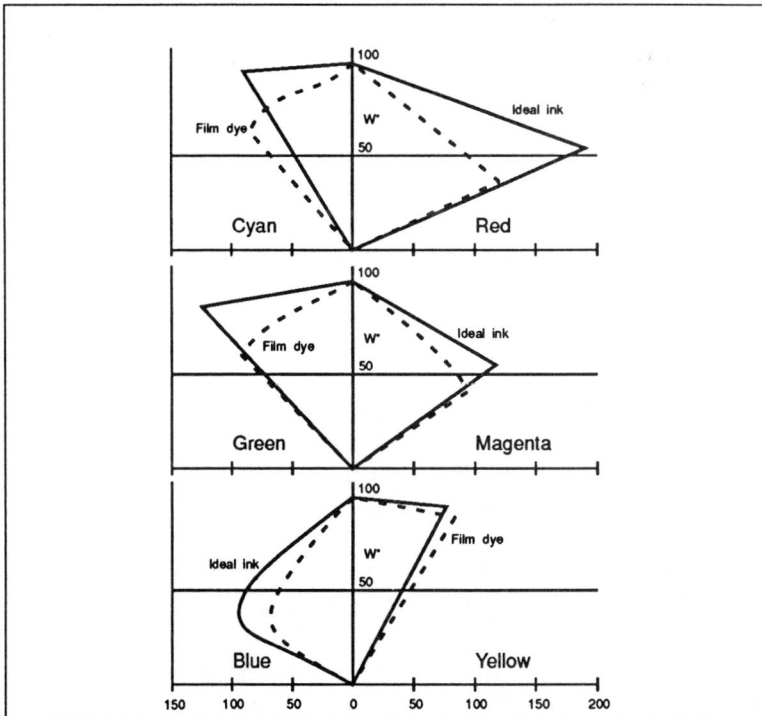

75

Studying these we now see that whilst the cyan does have a larger gamut than the ideal ink, this is only at lightness values below about 70%. (Hence the difficulty subtractive systems have in reproducing bright colours). Similarly we can see that at low lightness of yellow, magenta and green, the dyes have slightly greater gamuts.

If the same sort of plot was made of BS 4666 inks it is likely that they would show a greater gamut in the cyan and green at low lightness but would probably not compare so favourably as the dyes in the remaining four colours. Of course, this particular type of plot only shows the six primary and secondary colours. It could be extended to take in intermediate colours or alternatively have a different diagram of hue against W^*. However, we have considered enough to show the complexity of determining optimum ink colour.

The so-called ideal inks are in some ways an objective to aim for but can suffer a restricted gamut for darker colours. However, the darkening effect of the unwanted absorptions in real inks can give rise to problems in reproducing bright colours even when colour correction is applied.

The BS 4666 inks are far from being ideal inks. Inks could be produced much closer to the ideal, utilising far more expensive pigments. In practice, the bright colours which these would permit to be reproduced are rarely encountered and thus the extension of the gamut in the darker part of the colour space is frequently more valuable, so long as proper colour correction is undertaken to compensate for the hue shifts caused by the unwanted absorptions.

Thus we can see that the choice of inks for colour reproduction is not unique. A whole variety of cyan, yellow and magenta colorants could be used, each improving one part (or more) of the colour gamut at the expense of others.

The BS 4666 inks are a compromise between these factors and also that of price. However, if printers are frequently encountering non-reproducible colours, it would be instructive to carry out the sort of analysis described above and see whether a change in the colour of one or more of the inks would prove advantageous.

Of course, they should make sure during the analysis that the colours are truly non-reproducible and the problem is not caused by inadequate colour correction or insufficient ink density.

The alternative to changing the inks is to add extra colours to the reproduction. If we look back to Figure 4.8 it is readily apparent that adding another ink which has a colour defined by the plot x=0.15, y=0.7 will permit reproduction of saturated green colours not available from the process set. Similarly other colours could be added. The form of analysis described above may well show which are the best to use.

Spectral sensitivity of the separation system

Having decided that there is no unique optimum ink set, we now need to consider the filters used for colour separation and see whether this is any different, *ie* whether there is an optimum set of filters which may be used.

In fact we cannot consider the filters in isolation but rather the effect of the light/filter/lens/film sensitivity combination when photographic processes are being considered, and light/filter/ optical system/photo-sensor sensitivity when scanners are in use. A simple example of this is shown in Figure 4.10.

Fig. 4.10 Spectrograms of light sources and colour filters

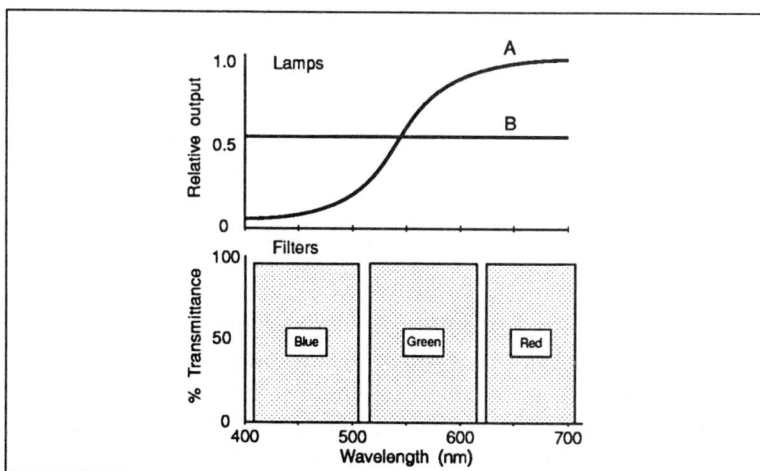

Here we show two different imaginary light sources, one with a high red and low blue output (A) and one with a uniform distribution (B). Also shown are three imaginary filters. When the different lights are used in combination with the filters, the relative amount of light transmitted at each wavelength differs, as shown in Figure 4.11. If the shapes of the combination spectral energy distributions are the same, but just differ in height, there is no problem. A simple variation in relative exposures will suffice to correct it but if, as in the case of Figure 4.11, the shapes differ, such correction will only be approximate. Obviously this sort of diagram can be extended to take account of the optical transmission and sensitivity. For the sake of simplicity we will just use the term filters, but it should be remembered that in reality the effect of light, lens and film also need to be considered.

Fig. 4.11 Combined spectrograms

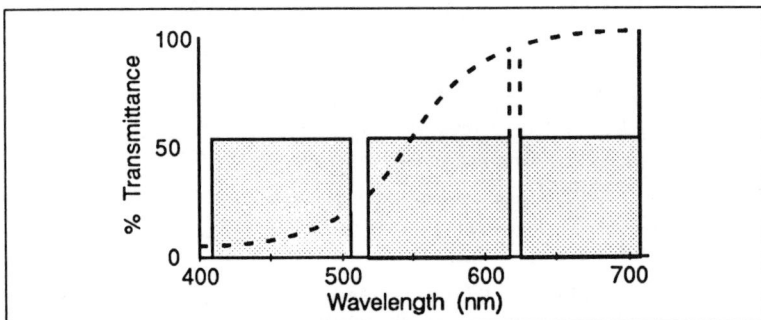

So what are the ideal filters for colour separation, if any? If we recall our initial discussion on colour matching it will be remembered that the fundamental property which permits colour reproduction by three inks to exist is that all colours can be matched visually by a mixture of the three colorants, if negative amounts of colour are permitted. Thus we can express graphically the amount of each ink required to match a particular wavelength of light.

This is shown in Figure 4.12 for the ideal inks discussed above. This tells us that if we are trying to match a monochromatic stimulus at 523nm, for example, we require zero units of the blue, 1.25 of the green and -0.75 units of red stimuli. This in turn means that we require a lot of yellow (to remove all the blue) and very little magenta (to pass a large amount of the green). But what about cyan?

Fig. 4.12 Constraints in reproducing a monochromatic cyan

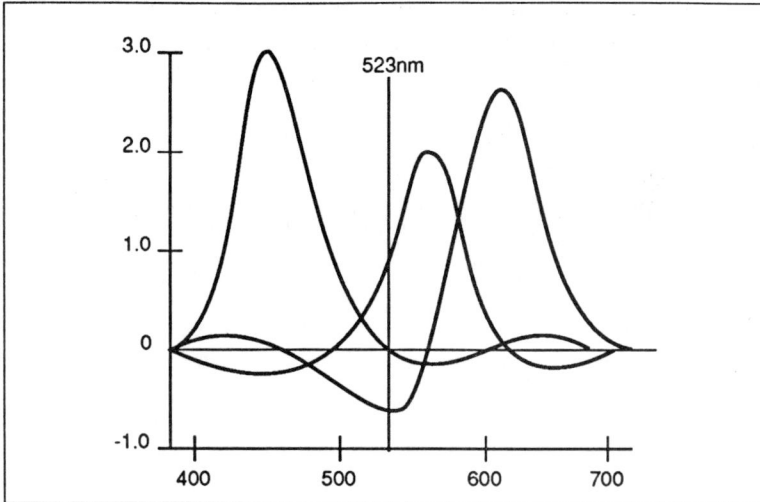

The fact that we need a negative amount of red means that we need more cyan than exists, obviously an impossible requirement. The best we can do is just to print the total amount of cyan available and leave it at that. This means that we are not reproducing a colour matching that of monochromatic radiation at 523nm, but rather a colour consisting of a mixture of 523nm and a small block of wavelengths approaching 700nm, which gives us the positive part of the red stimulus to offset the negative part of 523nm without affecting the green and blue stimuli much more significantly. Thus monochromatic light at a wavelength of 523nm falls outside the colour gamut of the ideal inks. In fact, we know this from an earlier discussion (Figure 4.8).

Since we are never (or at least very rarely) expected to match monochromatic stimuli by a reproduction process, we might ask what is the point of this. The reason is that *all* colours consist of a mixture of wavelengths, even those falling within the colour gamut of the ink set. The only reason that we can match the colours within the gamut is because the negative requirements at some wavelengths are more than offset by the positive requirements of others, giving a total positive requirement. Nevertheless, if the filter sensitivity of the separation system does not match the curves of Figure 4.12 (including the negative portions), even the colours within the gamut will have a colour error in their reproduction. At first glance this statement may appear confusing. Why should the filters need to match these curves?

We remember from the discussion on basic reproduction that a red filter is necessary to produce the cyan separation. This was so that the red light passed by the filter was recorded on the negative and thereby defined the amount of cyan ink necessary to absorb the appropriate amount of red light in the reproduction. Taking this argument to its limit, we could imagine using a set of filters each transmitting only a very narrow waveband (or even a single wavelength) and this would fulfil the requirement. This would only be the case if every colour in a group of originals which matched visually transmitted (or reflected) the same amount of light at the wavelengths defined by the filters.

Unfortunately, this is not the case. It is not a necessary requirement for colours that match to have the same reflectance or transmittance at any particular wavelength. This is due to metamerism. Thus two colours which match visually but do not emit equivalent amounts of light, as defined by the bandwidth of the filters, will be recorded differently on the film and hence differently on the reproduction. Thus, one or both colours will have a colour error.

A similar sort of argument, based not on metamerism but the relative amount of light reflected by different colours at different wavelengths within the same pictures, shows that colour errors will occur within a single reproduction. Thus we can see that only by using the colour mixture curves of Figure 4.12 can we hope to obtain accurate colour rendition.

Obviously it is impossible to produce filters with a negative transmittance and thus errors in colour reproduction are always present. In practice, the filters used frequently do not even approximate the positive parts of the colour mixture curves, and this tends to lead to even greater errors of the type described, so

80

we should try to establish why such filters are still commonly used.

The first point is that real inks do not have a clearly defined absorption band but, as noted earlier, absorption varies with dot size (or ink film thickness). This means that the perceived colour of the ink changes depending on the density level printed. Since the colour mixture curves of Figure 4.12 are defined by the absorption band for the ideal inks, it follows that there is no unique set of colour mixture curves for real inks.

A more serious problem with using filters matching colour mixture curves is associated with the difficulty of obtaining negative portions, and without these, significant desaturation of colours can occur together with errors of hue. By using filters of a narrower bandwidth, saturation is frequently increased. This can be understood by considering three filters, with no overlap, each primarily transmitting in the absorption band of the ink. Each separation will be a simple record of the amount of red, green or blue light emitted by the original, with no possibility of this same record being partially recorded on another separation and thereby desaturating the colour. However, as we have explained above, this will invariably lead to errors in the reproduction of colours, particularly where metameric effects are present. In fact this last consideration is largely irrelevant where colour correction is undertaken because the correction procedure has the effect of removing unwanted colours and thereby improving the saturation. Thus we should still conclude that a set of colour mixture curves would be preferable.

Now it is preferable to produce all-positive sets of colour mixture curves as we shall discuss later. However, there is some evidence that many of these give rise to high colour masking requirements which are impossible to realise in practice, because of their considerable overlap.

Hanson and Brewer, in 1955, concluded that is not important to use colour mixture curves, and this eases the masking requirement. Yule pointed out in 1967, however, that their data suggests that by doing so, errors are reduced by more than 50%.

The advent of many different dyes in photographic materials in recent years has increased the problems of metamerism, and it is certainly true that the attention of researchers is once again being turned to improving the filters in terms of their colour mixture capability. Quite what shape they should have is not certain, however, bearing in mind the variability in the absorption band of real inks as the dot size varies. It has been suggested that any set of colour mixture curves will suffice, and certainly this will ensure that metameric colours in the originals are reproduced alike, but it seems that this may have the effect of reproducing two metameric reds, for example, as a pair of matching oranges. To what extent the balance between separations can be adjusted to accommodate this is not completely clear.

From this discussion we can see that a number of compromises are present in the reproduction process. Apart from the fact that colours outside the gamut of the ink set are often encountered, we also have the problems associated with the low colorimetric quality of the separation system. It is for these reasons that printed reproductions frequently suffer from errors when compared to the original. A major difficulty is that it is still not agreed exactly what constitutes an optimum reproduction, given the limits of the process, but as this problem is solved and the separation filters are more nearly adjusted to match colour mixture curves (particularly in scanners) this problem should diminish.

Colour correction

We have looked in some detail at the inks used for colour printing and noted that relative to the ideal inks they have unwanted absorptions, *ie* they absorb light in regions of the spectrum where they should reflect. In a sense, then, each can be considered in these wavebands as behaving like one of the other inks. If we look at Figure 4.13 we see the spectral reflectance curves for the ideal inks and also for solid areas of real inks.

82

Fig. 4.13 Spectral reflectances of real and ideal inks

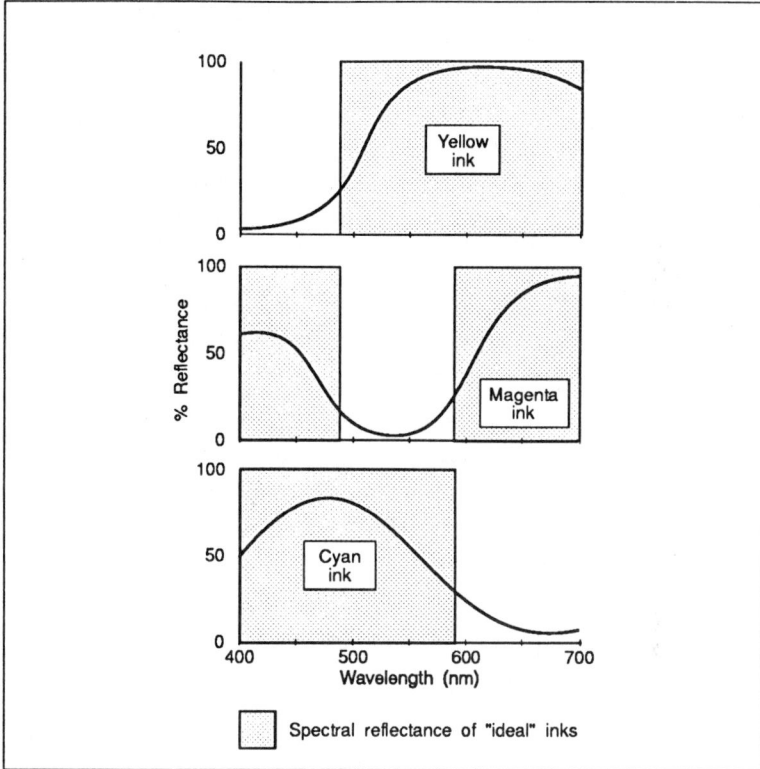

It is immediately clear that the magenta is particularly deficient in the blue part of the spectrum, and cyan in both green and blue. The yellow is fairly satisfactory. In an ideal masking system it is desirable to correct for all the unwanted absorptions and this means six corrections, two for each of the unwanted absorptions of each ink, but in practice concentrating on the major deficiencies mentioned above generally proves adequate. So it is apparent that we need to correct primarily for the blue absorption of magenta, and the green and blue absorption of cyan. Thus we can think of the magenta ink behaving as if it is contaminated with yellow (absorbing blue light) and the cyan as if it is contaminated with yellow and magenta (absorbing blue and green light). If we make separations with no correction for these effects, then areas

83

containing magenta ink will be too yellow and too dark, thereby looking yellow-brown, and areas containing cyan too dark and too red, looking brown. This is indeed the appearance of uncorrected reproductions, they do look dirty and brown.

When considering colour reproduction there is a natural emphasis on a consideration of the colours. It is often overlooked that many reproductions contain important areas of greys and greyish colours. The lack of purity of the primary inks (for that is what absorption errors amount to) not only means that secondary colours are not reproducible, but neither are tertiary colours (printed colours which contain all three primary pigments). In other words, when overprinting equal quantities of the three primary inks, a grey is not attained. This is important because grey is an easy colour to perceive in most reproductions. It is after all a 'dark white' and is readily comparable to white paper in terms of its colour. So lack of colour correction is readily perceived in neutral greys or greyish colours.

A considerable improvement can be made relatively simply by adjusting the balance of the colour separations in such a way that neutral greys are reproduced as neutral greys. This is achieved by determining the relevant dot sizes necessary to produce greys and adjusting the exposure of the separations to achieve this. To achieve this effect we would expect to reduce both magenta and yellow dot sizes relative to cyan, and yellow dot sizes relative to magenta, and thus compensate for the blue and green absorption of the cyan and the blue absorption of the magenta. This would remove the brown effect described above.

If such an adjustment is made it not only affects the neutral colours, but also the other colours in the picture and this improves the overall appearance. The result is not correct in all colours because the correction has been carried out based on equal dot areas (greys) and will not be sufficient, for example, in colours where the magenta dot is larger than the yellow because here the extra correction will be too great.

This procedure is known as obtaining grey balance. It is analogous to aligning the colour space of the original and the colour space of the reproduction by reference to the axis of the colour space rather than by attempting to make the peripheral colours match.

We shall see that in practice grey balance is achieved by BS 4666 inks when the yellow and magenta dot sizes are approximately equal and the cyan dot is larger, not with larger magenta and cyan as described above. This effect occurs because the inks are chosen to meet a requirement known as the balanced hue effect which, as we shall see, is very important when we come to full colour masking, as opposed to greys only.

If the green absorption of the cyan ink is greater than its blue absorption then it is possible to achieve the balance as described. The magenta ink is first reduced to a level where the unwanted green absorption of the cyan is accommodated by only reducing the yellow to the same level as magenta. For this to happen the blue-green absorption of both magenta and cyan inks must be in the correct ratios and it is when this occurs that the balanced hue requirement is met, as in BS 4666 inks. Of course these values are dependent upon ink film thickness and these need to be in balance as well, but we shall return to this point later.

We have established now a condition where the neutrals are being reproduced as neutrals but further correction is still required in certain colours which still tend to be too brown and too dark. These colours will be primarily those containing more magenta than yellow and more cyan than yellow and magenta (ie blue-greens, purples, maroons and some browns).

We also need to reduce the balance in some other colours with relatively high yellow and magenta content. In order to correct these it is necessary to undertake colour masking. A point needs to be made concerning the need for colour masking. It is often stated that one role of colour masking is to correct for the limitations of the filters. This is a fallacy.

So far all the colour masking theory developed above has been undertaken to correct for ink deficiencies. It has been assumed, though not stated, that the colours being reproduced and their reproductions are not metameric. In the simplest case this would mean that we are producing printed reproductions from *printed* originals or at least originals with dyes of precisely the same spectral characteristics as the ink.

If the filters used for separation and colour masking match colour mixture curves, even this criterion would not apply. We would only be correcting for ink deficiencies. If we take this a stage further and undertake separation with the set of colour mixture functions defined by the inks, even this would not be necessary, but as already pointed out there is no unique set of colour mixture curves for real inks.

Thus again we return to the fact that some research into this question of optimum separation filters is still required. Since we do not use filters having colour mixture characteristics for separation (because of the impossibility of producing the negative portion and possible difficulty in colour masking with all positive curves) it does become necessary to consider the problems of metamerism.

Metamerism occurs when two colours appear to be the same although they are known to be made differently. Their similarity will vary depending on the spectral composition of the viewing light (illuminant metamerism) or on the observers (observer metamerism). Metamerism is important in the printing industry because usually the original is a photographic dye construct and the printed reproduction is formed with ink pigments. If a pre-press proofing system not using inks is used in the process, metamerism is further complicated.

In fact it has been shown that metameric effects between original and reproduction require modifications to the colour masking procedure if colour mixture sensitivities are not used. However, it should be noted that what we are then correcting for is really a problem relating to the original colorants, not the filters.

In order to analyse these effects the analysis becomes somewhat complex and requires a degree of mathematical analysis beyond the scope of this discussion. For this reason we will largely confine our considerations to the non-metameric case. We should also note that photographic colour masking is done for purposes other than colour correction, particularly tone reproduction and undercolour removal. These should not be confused with the masks for colour correction, as their functions are quite different. In electronic systems, these functions are normally performed separately either by different circuitry or by different look-up tables.

We have now decided that we have a situation where the primary problems arising from the inks relate to unwanted blue absorption by the magenta and unwanted blue and green absorption by the cyan. This can be considered akin to the cyan ink behaving partly like the yellow and magenta inks, and the magenta ink behaving partly like the yellow ink.

In order to achieve full colour correction we need to extend the masking selectivity where the grey balance correction is inadequate, *ie* in colours containing more magenta than yellow, for example, and decrease it selectively where the grey balance correction is too great as in the inverse case. In fact it is convenient to consider colour masking prior to obtaining grey balance and impose this later as we will see.

Whilst there are various practical methods of undertaking masking, only the simplest, the positive masking method, will be described here. Two stage masking, tri-packs, *etc*, are only extensions of this method with certain advantages and disadvantages.

We shall start by considering the problem of the blue absorption of the magenta ink. In order to correct for this we need to reduce the blue absorbing ink (yellow) in proportion to the amount of magenta printed. In order to achieve this we produce a positive from the magenta separation negative and bind it up with the yellow separation negative.

The density of the positive is produced to a level corresponding to the hypothetical amount by which we are imagining the magenta to be contaminated by yellow. That is, the greater the unwanted blue absorption of the magenta ink, the more yellow we need to remove and hence the greater we need the density of the positive mask to be.

By binding the yellow negative and magenta positive together we have a situation where, for example, the clear areas of the negative (which will produce a 100% dot in the yellow plate) have a density added to them proportional to any magenta printing in that area. The resultant dot area of the yellow is thus reduced on the plate.

Similarly, to correct for the unwanted green absorption of cyan we bind a positive produced from the cyan separation negative to the magenta separation negative to reduce magenta wherever cyan is printing.

However, what about blue absorption of the cyan ink? Thus far we have not corrected for it and yet commented earlier that it was a fairly serious deficiency.

The answer lies in the balanced hue requirement discussed earlier. Since the magenta mask for yellow was made from the uncorrected magenta separation, the mask had not been corrected for the unwanted cyan and thus effectively acted as a record of the magenta plus some cyan.

It can be shown that if the balanced hue requirement is met, the effective cyan correction applied by the magenta mask is just that needed, and a separate mask to correct the yellow for the blue absorption of cyan is not necessary. If inks deviate seriously from this requirement it is necessary to produce a magenta mask for yellow made from the corrected magenta separation and also a mask for yellow made from the cyan separation and bind both of these to the yellow negative.

The resultant flow diagram for separation, resulting from the application of masking, is shown in Figure 4.14.

Fig. 4.14 Colour separations with a masking stage

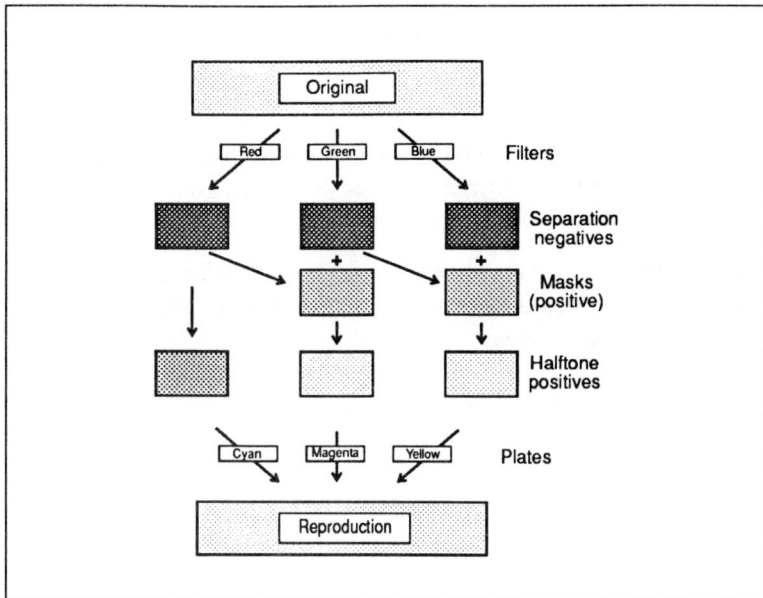

After masking has been carried out we need to apply the grey balance condition considered earlier. This acts as the necessary base to bring the balance of the separations together again.

Whilst greys have been corrected by masking (contrary to popular belief) it is necessary to ensure that the grey balance condition is met in order to ensure that the separations are in balance.

If the masking has been properly carried out and the grey balance condition achieved then all reproducible colours within the reproduction will be accurate (subject to the problems briefly discussed below).

As already stated, the positive masking method described above is not the only one which may be used (in practice it rarely is), but it is the simplest one to describe. All other methods have a similar effect, but obtain it by different routes.

One alternative is to make negative masks, not from the separations but direct from the original. This has an advantage in that the filters for mask production may be different to those for separation and can be designed to correct for the small unwanted absorptions not considered in the positive method above. Thus, slightly improved correction can be achieved.

In fact the positive method can also accommodate these but account has to be taken of interaction between the masks, as described for the inks not having a balanced hue, and the procedure becomes rather complex. Nevertheless the positive system can produce very good results without this.

Apart from incomplete correction for all the unwanted absorptions, masking does suffer from other limitations. The theory developed above assumes that ink densities are both additive and obey laws of proportionality. The former means that if the blue absorption of the magenta and the blue absorption of the yellow each give rise to a specific density then the total blue absorption, if the two are overprinted, is defined as being equal to the sum of the densities. The latter means that despite variation in concentration (or dot area) of a single ink, the red, green and blue densities should remain in proportion to one another.

In practice these laws are only obeyed approximately, for a whole variety of reasons, and deviations depend upon the ink/paper/measuring technique combination. Deviations from these laws necessitate what is known as non-linear masking, *ie* the masks do not have relationships which are linear as the density is increased.

A graphical description of a non-linear mask shows not a straight line as density increases but a curve. This means that at certain densities it is necessary to apply more or less correction than predicted by the simple consideration of the unwanted absorption bands of the single inks. Using a linear mask will give perfectly adequate results for many subjects, but some colours will not be correctly reproduced. The fact that a variation in masking is required can in fact be easily predicted from Figure 4.7, where we see that the absorption band changes as the dot area is altered. This almost certainly will affect proportionality, although it does depend upon the filters used. In fact the major problem associated with masking is the proportionality failure attributable to this cause, and this generally requires a higher degree of correction in lighter tones.

Nevertheless, proper correction can only be established by first basing it upon the unwanted absorption of the solid inks as described above and then introducing non-linearities to improve these.

In scanners, of course, masking is done electronically. In order to achieve this the so-called masking equations are used. These are a mathematical description of the procedure above, still assuming additivity and proportionality. In order to avoid the difficulties this causes it has proved necessary to have different equations for different areas of colour space, but even this is only approximate. That is why most scanners need so many colour correction controls, to enable the operator to compensate for the errors as appropriate.

Grey balance and tone reproduction

There can be little doubt that the most common causes of unacceptable reproduction are inappropriate tone reproduction and poor grey balance. Colour correction is important but the achievement of correct grey balance will go a long way to producing good colour rendition. Tone reproduction is probably the most crucial aspect since this affects whether a reproduction appears 'muddy' or 'dirty' or too bright. The problem with most poor reproductions which are submitted to Pira for investigation is tone reproduction inaccuracy.

Grey balance has already been discussed in some detail but is worthy of a little expansion. We are referring to the balance in neutral colours and assuming that neutrals in the original need to be reproduced as neutrals in the reproduction.

We have shown that differences in dot size between the various separations are necessary to produce greys because of the unwanted absorptions of the inks. In fact this is only part of the story. The ink film thickness at which the inks are printed also has an effect. For example, it would be possible, despite the unwanted absorptions, to achieve good blacks simply by adjusting ink film thickness and strength, and printing solid films of such colour. This is often done in the gravure process in fact. Adjustment of relative dot area between colours would still be

91

needed at lighter tones, however, due to the unwanted absorptions, primarily because of additivity and proportionality failure.

Generally, three solid colours overprinted produce a brownish black. To increase the cyan further to avoid this would result in much too heavy an ink weight and produce practical press difficulties. Alternatively the magenta and yellow could be reduced but this would result in poor reds and yellows. In practice the grey balance condition should be established with reasonable ink weights, cyan being the heaviest.

The maximum three colour density obtained by overprinting these solids will be slightly brown, but this will be largely corrected by the black printer. The grey balance condition at lighter tones is then established at these ink weights by use of a colour chart such as the Rochester Institute of Technology tone reproduction and neutral determination chart (TRAND) or similar.

This test form prints five different dot areas of cyan, from 16% to solid, and overprints these with different amounts of yellow and magenta. The grey at each dot area of cyan is determined visually and this defines the imbalance required to reproduce neutrals. This ratio is then applied after colour correction has been carried out. The full method is covered in Chapter 8.

Tone reproduction

As already stated, tone reproduction is undoubtedly the most important property of good colour reproduction and yet is probably the most complex.

The concept of tone reproduction in a monochrome print is simple to understand (although rarely achieved satisfactorily in practice) and so we shall start by discussing this. Basically tone reproduction is defined as the relative difference between levels of greys (brightness) on the original and in the reproduction.

If this relationship is not correct then highlight areas may appear dirty or too bright and shadow detail may be lost in a black mass. Precisely what this relationship should be is a matter of some debate, but in Pira report PR 143 (1979) the hypothesis is put forward that in general the tone reproduction requirement is linear, on a uniform visual scale, since this ensures that different magnitudes in the original are maintained in the reproduction. The perceived tone reproduction is very dependent upon viewing conditions, particularly the intensity of the viewing light and the density and colour of the area immediately surrounding the viewed piece.

If the contrast range of the original and the reproduction, and the illumination level for viewing both are identical, then an exact reproduction of tones can be made. In practice, the printing process has some difficulty in achieving the contrast range of the photographic process.

The term 'tone reproduction' has been taken here to be simply the relationship between input copy and output reproduction. However, the term can be applied to any stage of the process, *eg* input copy to screened positives. The sum of all these separate stages can be described in a quadrant, or Lloyd Jones diagram (see Chapter 9). However, in order to construct such a chart, it is necessary to start by defining the input/reproduction relationship.

In previous work by A J Johnson (1977) on tone reproduction, Pira has developed the work of Bartleson and Brennemen where the idea of a linear compression is applied not to density, but to luminosity. Relative luminosity (relative brightness in the US), can be defined as the perceived brightness of a point in a scene relative to the apparent brightness of the whitest point in the scene. The term derives from the notion that the amount of light reflected from an area is meaningful objectively, but that when viewing the scene containing the area, the eye adapts to the scene in general and the whitest point of the scene in particular. So the perception of the reflected light changes according to the adaptation that occurs. Thus luminosity expresses the subjective effect of a stimulus. We can then consider the linear compression of the luminosities of the original, to the luminosities of the printing process. Since the luminosity is a subjectively based function, there would seem to be little future in it. Subjectivity is, after all, an intangible. What this approach attempts, however, is to provide a conversion from a subjective perception to a measurable entity – density.

Since a perceptual expression must take into account all the factors that affect perception, it is necessarily complex. The mathematics expresses a power function, with a decay, and contains parameters to express the illuminance and surround luminance. If the illuminance or the surround luminance change, then the parameters change.

It is necessary to define the luminosity *versus* density function for each differing viewing condition, so that if a transparency is to be compared with a printed reproduction, both must be converted to luminosity according to their own specific viewing conditions. Figure 4.15 shows how density is converted to luminosity in the first place.

Fig. 4.15 Density to relative luminosity for two viewing conditions

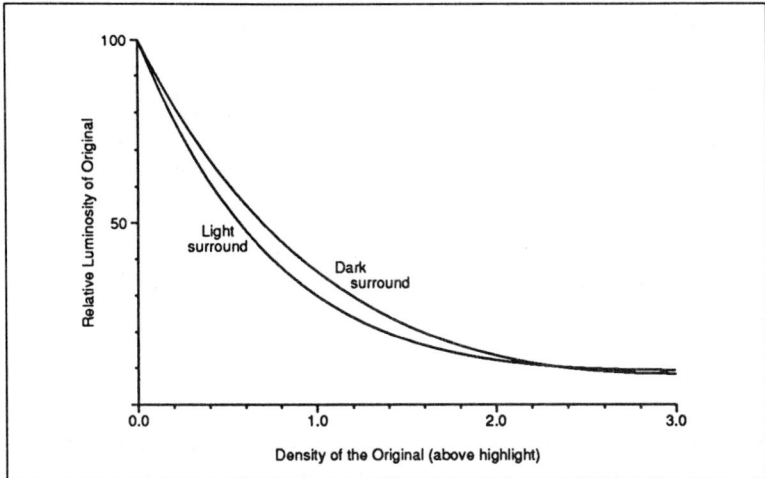

Figure 4.16 shows two viewing conditions, and the arrangement that allows the linear compression between luminosities to take place, but with the resultant compression expressed in terms of density.

94

Fig. 4.16 Tone reproduction accomplished in appearance terms

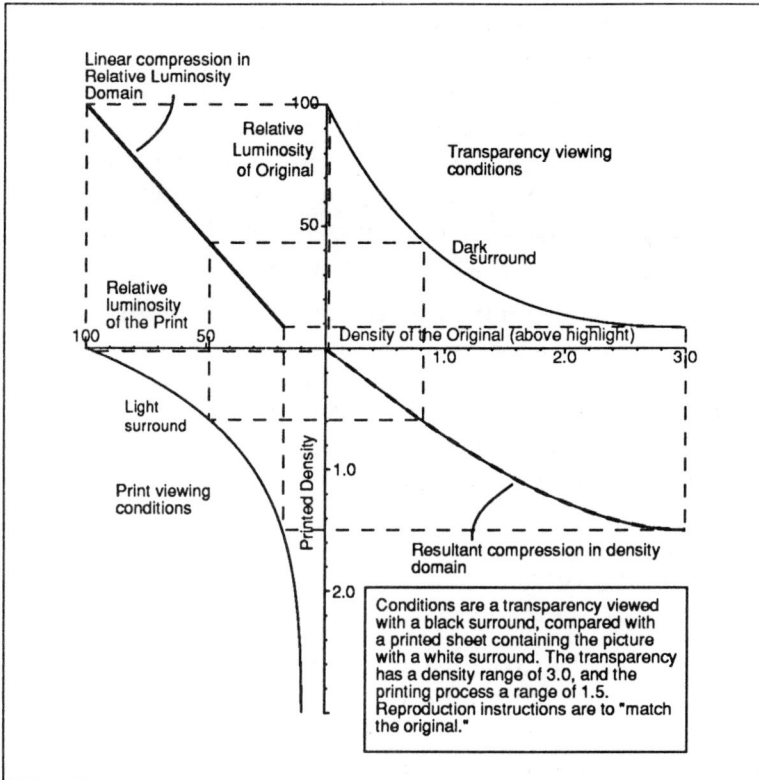

Linear compression in
Relative Luminosity
Domain

Relative
Luminosity
of Original

Transparency viewing
conditions

Dark surround

Relative
luminosity
of the Print

Density of the Original (above highlight)

1.0   2.0   3.0

Light
surround

Printed Density

Print viewing
conditions

1.0

Resultant compression in density
domain

2.0

Conditions are a transparency viewed
with a black surround, compared with
a printed sheet containing the picture
with a white surround. The transparency
has a density range of 3.0, and the
printing process a range of 1.5.
Reproduction instructions are to "match
the original."

This type of compression is in use on a commercial scale, having
been incorporated in a series of colour scanners available interna-
tionally. Its success relies on the implicit understanding that the
reproduction is required to compare visually with the original.

The point is that if the idea of a linear compression of luminosities
is valid, and it is, then it must be emphasised that the viewing
conditions for both the original and the reproduction must be
stated in advance. In most cases, an acceptable assumption is that
the reproduction will be viewed in light surround conditions (a
picture on white paper). If the picture will be surrounded by a
printed black border, this is usually known in advance of the
reproduction process, and so can be taken into account. If the

transparency is designed to be projected, as a 35mm transparency is, then it can be assumed it should be viewed both with a dark surround, and with an allowance made for flare light. (The material is processed to render a gradient of 1.5 for this reason.) If the original is a photographic print, the density to luminosity conversion must be that representing light surround viewing conditions. Once stated the viewing condition cannot be changed. If therefore it is stated, or assumed, that the original will be viewed with a light surround it cannot be masked down at proof passing time in order to give a 'better' picture.

The surround in this case means the immediate surround of the picture, but this still leaves the question of the surround in the fuller sense, that is the 'background' of the viewing area. The eye after all can cover a visual field approaching 180° and the original or the press sheet are unlikely to extend to the edge of the visual field. Remembering that 90% of scenes integrate to an effective neutral grey of 18% reflectance, it should be arranged that reproductions and originals are viewed in these conditions. The adoption of viewing conditions conforming to ISO 3664 is recommended.

When we consider colour the subject becomes somewhat more complex but not dramatically so. Firstly we should note that the reproduction requirement for a scale of greys in a colour reproduction is precisely the same as for a monochrome print and thus the linear visual relationship discussed above is still equally valid.

The relationship between tone reproduction, grey balance and colour correction is a very interesting area because we introduce the problems associated with hue and saturation of colours as well as brightness. So long as we restrict our attention to a grey scale and assume good grey balance is achieved then the tone reproduction is easy to define, but this reproduction will also affect the reproduction of colours.

Let us demonstrate this by a simple example. Suppose that the ideal relationship for a reproduction is a straight line, but that all the separations are produced in balance, and with a reproduction characteristic bowed upwards, as in Figure 4.17.

Fig. 4.17 Luminosities of original and reproduction

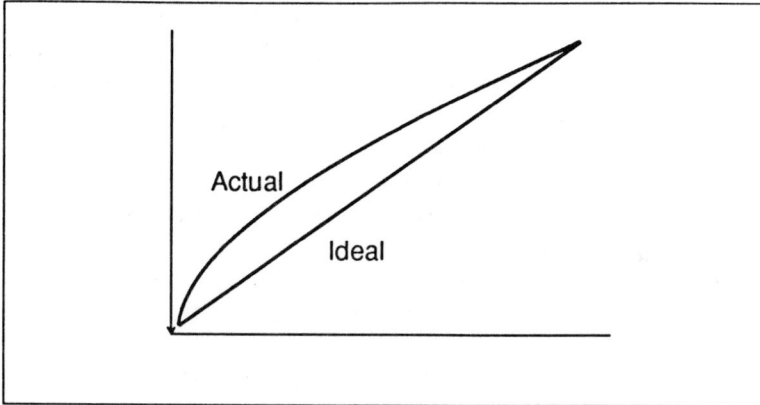

Any colour within the reproduction with more or less equal dot areas (*ie* near neutrals) will be reproduced with the correct hue, (because the balance is correct) although the saturation and brightness (strength) will be in error.

However, consider a colour which should be reproduced with a 50% dot of one colour and 100% dot of another, an orange for example. Assuming the colour corrections are correct then the linear reproduction will give the correct result. If the bowed reproduction characteristic is used, however, the result will be an increase in the 50% dot but no change in the 100% (it can go no higher) and thus an orange will tend to red as the proportion of magenta to yellow increases.

This is why tone reproduction is so important in colour printing. Not only does it define the neutral scale, in terms of brightness, but also the colours, in terms of brightness, saturation and hue. It is impossible to define a colour correction system without first defining the tone rendition and grey balance. Of course accurate tone reproduction and grey balance do not in themselves guarantee good colour reproduction – if the colour correction is inaccurate some other colours will not be accurate – but they go a long way towards it. This is why we generally stress the importance of tone reproduction.

The statement made here that colour correction is affected by tone reproduction is an interesting one, since if we consider the assumptions made previously it was assumed that colour correction was determined solely by the unwanted absorptions of the solid inks and thus should not be affected by tone reproduction. How do we reconcile these two statements? The first point to note is that even with the simple colour masking theory previously described it is essential that the masks have the same tone reproduction characteristics as the separations, although at the required contrast. If this is not the case it is immediately apparent that the correction will be inaccurate. Remember that the mask should be a record of the amount of one of the colorants to be applied, and in tonal values this obviously depends upon the tone reproduction applied. Thus the mask characteristics must accommodate this. The other problem rests on the fact that masking is not in general linearly related to the solid ink unwanted absorptions because of additivity and proportionality failure. Thus the deviations from linearity introduced will be very dependent upon tone reproduction. It is for these reasons that colour correction must be adapted to suit tone reproduction.

A useful way of checking for accuracy of colour correction for a given tone reproduction characteristic is by means of a colorimeter. If, for example, we see from this that reds are consistently produced too light, whereas greys are in balance, then it indicates that the colour correction is faulty, perhaps giving rise to too little magenta or cyan in reds. The exact type of deviation will give the precise reason.

The black printer

All our discussions so far have been largely concerned with the principles of three-colour reproduction. In practice, however, we also add black. This does not in any way affect the theory so far but instead adds a complicating factor. The primary reason for adding black is because of the limited maximum density achieved with the three colours above. This is exacerbated by the application of grey balance. Grey balance calls, in nearly all cases, for the reduction of both yellow and magenta from greys and any reduction of ink must reduce the density of the printed colour. This further reduces the printed contrast because 'greys' includes blacks in the picture. Thus the addition of black ink is an attempt to compensate for this. Most printed works which include pictures and words require that the words are printed in black ink – so the black is available.

Once black is included in the reproduction, other uses may be made of it. By means of a technique known as undercolour removal (UCR), it is possible to replace proportions of the three process inks, in the neutral and near neutral areas of an image, with one ink – the black. UCR is the reduction of cyan, yellow and magenta dot areas, in correct proportion to one another as determined by the grey balance characteristic where all these are present, and printing the appropriate amount of black instead. Grey component replacement (GCR) is a further implementation of the principle, reducing the grey component from all colours in a reproduction (not just the neutrals), and replacing them with black ink. GCR is known under a variety of acronyms depending on the source. The most ludicrous of these is 'achromatic reproduction' which means literally 'no-colour reproduction'.

Both these techniques help in avoiding drying difficulties associated with piling four colours on a sheet whilst all are wet, and make control of the printing process less difficult since it becomes less sensitive to changes in balance between the colours. This is particularly useful in reproducing originals containing large areas of near neutral colour.

The addition of black makes the theory and practice of colour reproduction somewhat more complex because for the colours in a reproduction, there is only one unique mix of three (primary) colours that will achieve accurate colour, saturation and lightness. When the black is introduced there is no unique mixture of four inks which will reproduce them. A grey colour can be reproduced by three colours only, by black only, or by a mixture of three colours plus black.

However, it is essential to think of a reproduction in terms of three colour mixture to obtain correct hue and saturation rendition of colours. The black is added firstly to achieve optimum rendition of darkness and secondly considered for use in replacing the process inks in greys or greyish colours. If this not done properly, adding black can seriously detract from a reproduction.

Chapter 5 *The colour scanner*

Colour reproduction in printing has been practised for most of this century by chromo-lithography, and letterpress and gravure etching techniques. Both the colour scanner and lithography have increased in importance, worldwide, over the past 40 years. Given the requirements for the colour reproduction system in Chapter 4 we can consider how these may be satisfied.

Colour reproduction by photographic means has been success-fully employed, and its implementation has been considered above, particularly with regard to the necessary colour correction masks. Such methods are constrained by the characteristics of the photographic materials used, *ie* the contrast or gamma of the material. The operating costs of a photographic system are necessarily very high. It was not unusual to produce somewhere between four and 12 films prior to the actual halftone colour separation films that are used to make the printing plate. Each film or set of films has to be exposed, processed, washed and dried before use, so that the execution time of the pre-press processes can be measured in days.

The colour scanner has shortened this lead time dramatically. The scanner has progressively taken more of the necessary processes into the electronic domain. It is possible now to implement filmless repro from the original to the printing surface. In filmless repro, the image is input to a computer, stored and output at a later time. It is not, therefore, direct from the original to the printing surface.

Types of input scanner

The first generally available scanner was the Vario-Klischograph (see Figure 5.1), manufactured by Rudolf Hell in Germany. It was capable of engraving letterpress blocks direct from the original transparency. It was able to enlarge or reduce the size of the reproduction, since the printing surface, or block, must contain the picture at its required size. The enlargement was accomplished by means of a pantograph arrangement, a series of levers and bearings, where the relative distances traversed by the original and the printing block could be changed. The input and output beds attached to the pantograph were made to oscillate such that at the input a line of the original was scanned by three filtered photomultipliers. The signals from the photomultipliers were processed and the resultant signals used to control an engraving stylus which bore upon the surface of a copper plate. The stylus engraved at a rate determined by the required screen ruling and both input and output beds could be rotated to set the screen angle.

The Vario-Klischograph was a heavy, cast iron structure relying for its precision on the accuracy of machined sliding parts. It was expensive to manufacture and to maintain. The colour processor, a thermionic valve analogue computer, was updated continuously.

Fig. 5.1 Vario-Klischograph engraving scanner

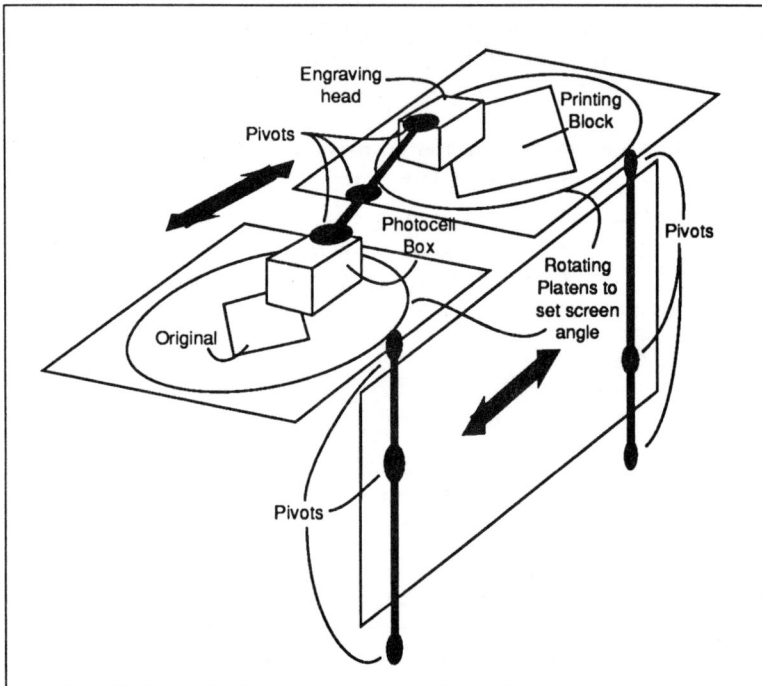

Modern scanners are based on solid state electronics with the scanning arrangement being either:

A rotary drum.

A flat bed – either the bed or the light source/detector assembly moves.

At present the greatest mechanical precision is obtained with the rotating drum arrangement.

Rotating drum

The rotating drum scanner is based on the engineer's lathe (see Figure 5.2). The lathe layout ensures precision which is necessary if dividing and reassembling the picture. The scanner consists of two main parts, the analyse section and the expose section. Each has its own drum, the analyse drum carrying the original and the expose drum carrying the film to be exposed with the colour separations. The rotating drums are usually mounted on the same axle and rotated together. However, the input and output units can be housed in separate units and linked by synchronising electronics. Photomultipliers are used to 'view' the original through the primary filters – red, green, and blue.

Fig. 5.2 Rotating cylinder or drum scanner

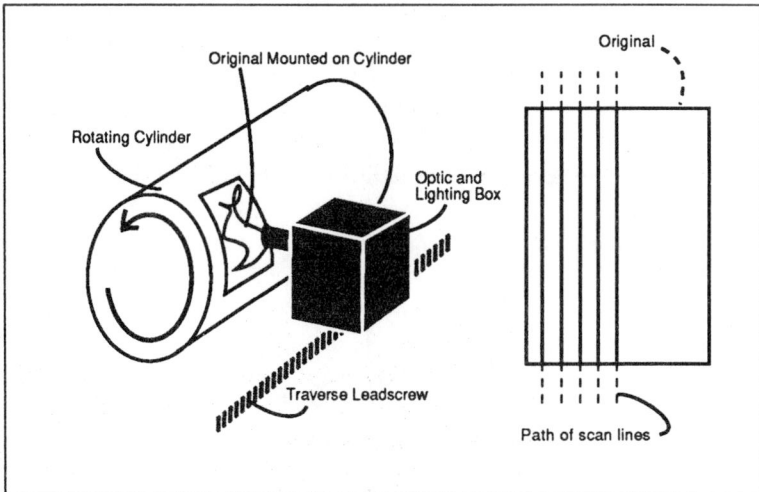

103

The path of the scan line on the original is actually helical in this design. If the original is held upright, the scan lines should ideally be square to the horizontal of the original. The pattern of lines is at a small angle from the ideal 90° but the error can be ignored for practical purposes.

The scan lines are formed from successive revolutions of the scanner drum. All the time the drum is rotating, the photomultipliers are sampling light from the original. The signals from them form a continuous electrical analogue of the densities of the original. In practice the signals are 'clocked' to form discrete samples called picture elements, or 'pixels' for short. Pixels provide an ideal format for the digitisation and storage of the image in computer memory.

Electronics

The photomultiplier signals are processed and form the exposing signal. It is the nature of processing that takes place which will be addressed. The discussion on colour masking in Chapter 4 forms the basis for what is required in such processing. Where in the photographic domain we are concerned with whether a positive or negative image is being considered, in the electronic domain positive and negative are relatively simple inversions one from the other. The following illustrations are centred on an electronic device known as the operational amplifier (op-amp for short) as shown in Figure 5.3. This is the basic currency of the integrated circuit. Whilst many of the integrated circuits (ICs or chips) in use are now significantly more sophisticated than the op-amp, it still provides a clear example of what is electronically possible.

Fig. 5.3 The basic operational amplifier

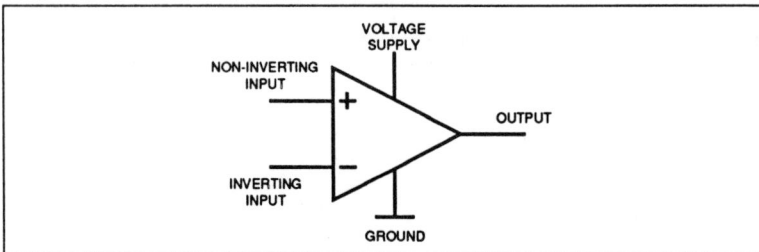

A close analogy of photographic masking techniques can be made by suitably connecting op-amps. For a positive process the non-inverting input can be used whilst for a negative process the inverting input can be connected. In this way both positive and negative signals can exist at the same time although at different op-amps.

Fig. 5.4 Configurations of operational amplifiers

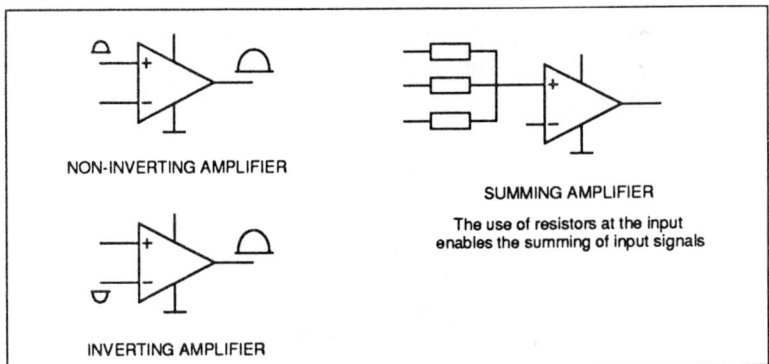

NON-INVERTING AMPLIFIER

SUMMING AMPLIFIER

The use of resistors at the input
enables the summing of input signals

INVERTING AMPLIFIER

Not so obvious, but equally important, is the idea of comparing two signals in an op-amp. In this configuration the op-amp will amplify the difference between two input signals. 'Amplify' is the correct term here but it should be remembered that the amplification factor can be any number between zero (nothing happens) to infinity (the op-amp acts as a switch, either fully on when a signal is present at the input, or off if no signal is present). A surprising number of op-amp configurations operate with a gain of 1 (the output is equal to the input but may be inverted).

Fig. 5.5 The operational amplifier configured as a comparator

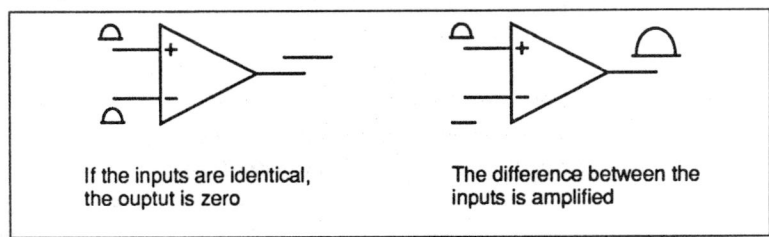

If the inputs are identical,
the ouptut is zero

The difference between the
inputs is amplified

As an example of what can be achieved in the electronic domain consider the following circuits. Using a series of op-amps (as shown in Figure 5.6) we can implement a form of single overlay masking described earlier for film work.

Fig. 5.6 Single overlay colour correction masking using operational amplifiers

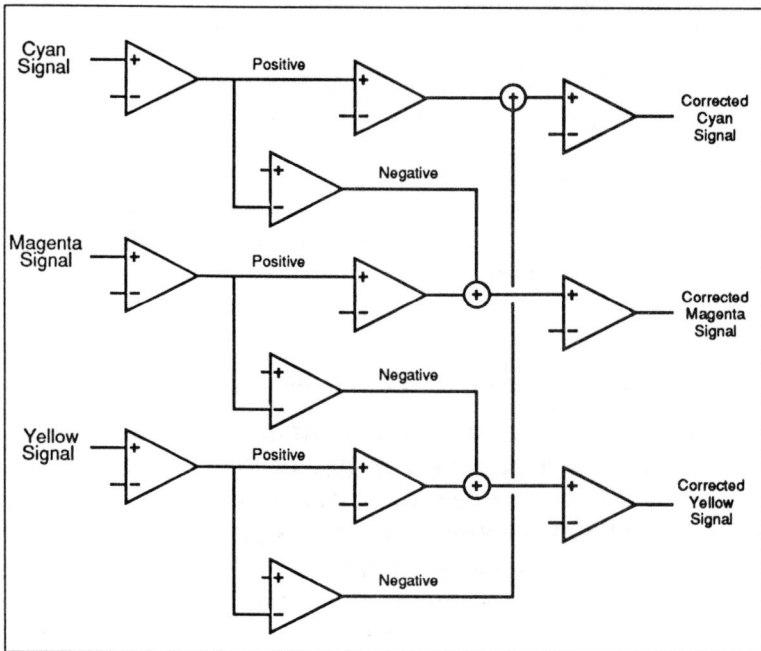

The superior double overlay masking technique can also be implemented (see Figure 5.7). When undertaken by photographic methods the double overlay masking system is expensive in both time and film, but it affords much control over colour correction. It is also possible, by changing the contrast of the masks, to implement colour transformations and extractions for producing special (non-primary) colour separations. The electronic equivalent of changing contrast is to adjust the gain of the op-amp.

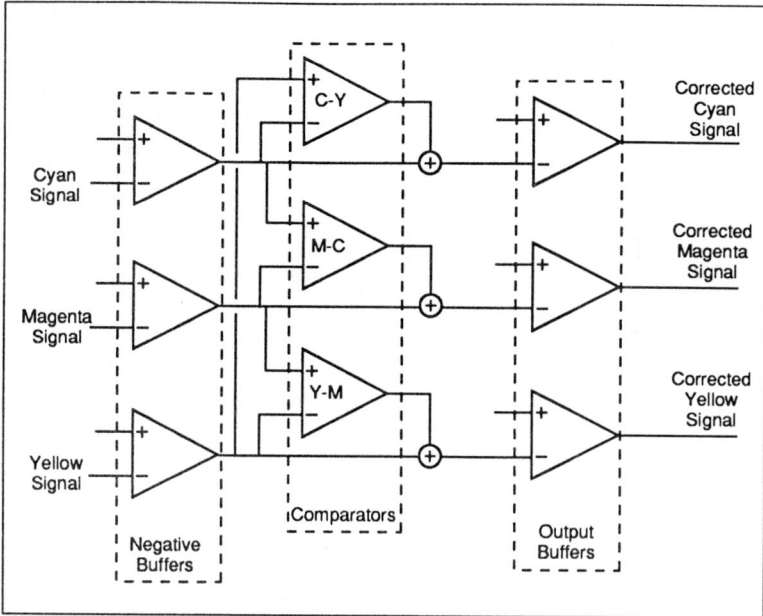

Fig. 5.7 Double overlay masking performed by operational amplifiers (much simplified)

The circuits above can be configured into the inner workings of a scanner and many scanners used these and similar techniques to implement the colour separation process. These, so called analogue, scanners were typified by their abundance of knobs and complex operator control panels. The operator had to set up each separation on the basis of the densities and appearance of the original and to ensure that the scanner as a whole was properly set up and calibrated. In the mid-1970s, when such technologies were prevalent, the cost of each op-amp was about one third the cost of a sheet of average size separation film. Of course when a whole scanner was considered, which may contain many thousands of op-amps, the cost was far from cheap.

Modern scanners increasingly use the digital computer. In a digital design the computer, or computers, are programmed with the mathematics of the colour reproduction process. The photosensor signal is digitised, and the operation of the scanner becomes a data processing operation.

Flat field colour scanners

The 'discrete points on a line' approach to scanning is giving way to the idea of scanning a whole line at a time. For this to happen requires that the scanner be configured with a full line of photo-sensors which 'see' a line of the picture at once. The development of the charge coupled device (CCD) array has enabled this form of scanning to be considered, and it is set to become increasingly important in the future.

The CCD array is essentially a line of capacitors formed on a semi-conducting substrate in the same way as integrated circuits are constructed. The device was developed as an electronic, analogue, short term memory, or shift register, for use in sound reproduction systems. The idea was to introduce reverberation by mixing a time delayed signal with the real time signal.

Each capacitor is essentially a semi-conducting photo-diode whose junction is designed so that light falling on it is stored as an electrical charge. In action the device is exposed to light in one phase, then, in a second phase, a signal causes the device to transfer that charge to the next device in line. A number of signals later the charge appears at the output and is passed on for further treatment. In a linear array of CCDs the scanning is accomplished by electronic means in one dimension. It is necessary to move the image across the linear array in the orthogonal direction to achieve the two dimensional scan required.

Fig. 5.8 Detail of the CCD line camera used in the flat bed scanner

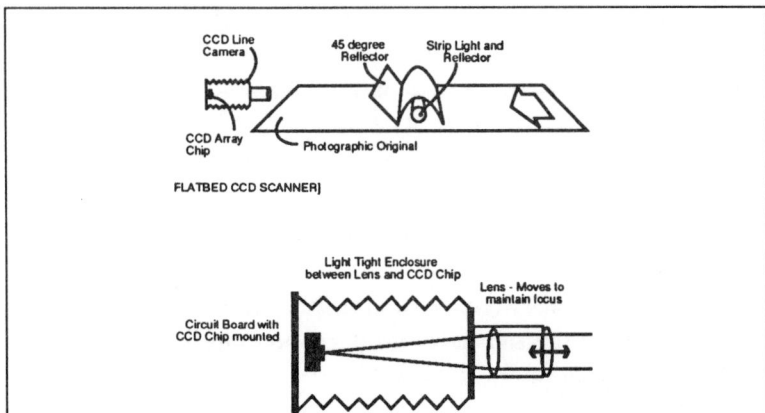

108

It would, of course, be desirable to arrange the CCD as a two dimensional array and carry out scanning as a wholly electronic process. This is done in video cameras. In moving pictures the image quality is far less significant than in still pictures where the picture is subject to closer scrutiny. The problem with CCD technology is that the construction of many components at the same time results in differences in behaviour of the finished photo-diodes. The production yield of working arrays increases if the number of devices per chip is reduced, hence the use of linear arrays as opposed to two dimensional arrays.

The use of CCD scanners in the reproduction area is somewhat limited by the quality of image that can be produced. The compromises on image quality come from the:

> *Limited dynamic range (contrast range).*

> *Effective resolution, limited by the density of packing of the individual photo-sites.*

> *Differences of photographic/electronic characteristics of each photo-site.*

> *'Noise' generated by the photo-sites and other electronic devices in the chip.*

> *Optical system necessary to focus the image on the device.*

> *Lack of precision in the moving parts of the flat field scanner.*

Many of these problems are temporary: the dynamic range can be extended by cooling the device and this is likely to reduce generated noise; the effective resolution may be overcome by optical scaling or by increasing the number of photo-sites per array; the differences in behaviour by improved production techniques; and the lack of mechanical precision by improved construction. All of these solutions will add to the cost of the flat field scanner which is its chief attraction currently.

Of course if all of the problems were overcome, the construction of a two dimensional array of suitable quality becomes possible and the need for mechanical scanning is no more. In other words a flat field camera can be used, where the 'scanning' is carried out electronically.

Unsharp masking

It should realised that however the original is scanned, whether by rotating drum scanner or by CCD flat field scanner, the image is effectively being captured whilst moving. Even given a two dimensional scanning array such as a video type of camera, the image contains compromises which affect its sharpness.

The solution to the problem is 'unsharp masking' which, despite its name, is used to enhance the sharpness of the image. The name comes from techniques used in film masking methods discussed in Chapter 4. If a mask is used it often reduces the contrast of the resultant image. If, however, the mask is made unsharp, that is diffuse, it will not resolve the fine detail in the original image. Therefore, in areas of fine detail, the mask cannot reduce the local contrast. These areas retain the original contrast and, as a result, appear sharper.

The provision of electronic, unsharp masking is an essential requirement on a scanner and was originally provided by the application of a separate scanning aperture.

Fig. 5.9 The apertures through which the original is 'seen' by the scanner

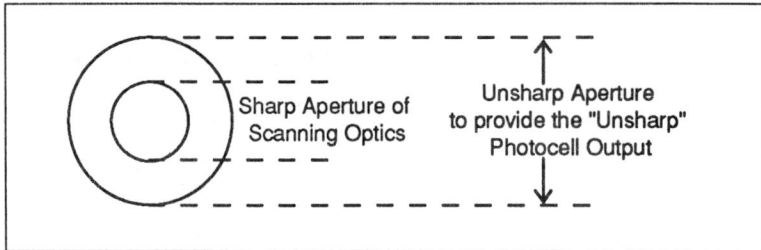

The signals from the two apertures were combined in the form of the 'sum of the difference' between them and resulted in a restored sharpness composite signal.

Fig. 5.10 USM signal generation

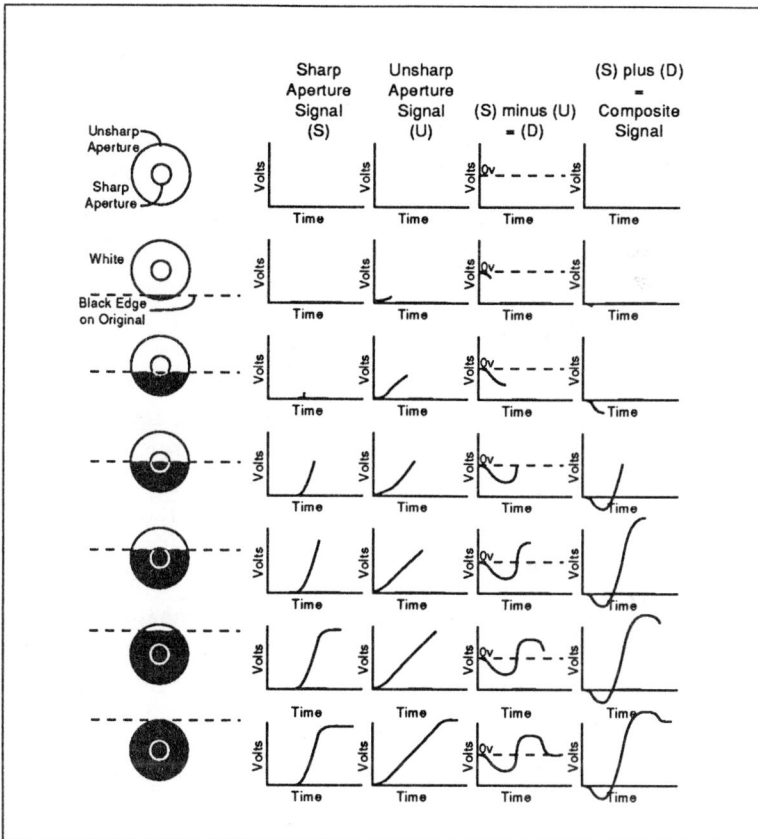

In the course of time, much greater use has been made of the unsharp signal resulting in the super-sharp images seen in all areas of print. Printed images can appear too sharp by the over-use of unsharp masking, containing hideous 'fringes' around the component parts of the picture. The natural limit to the amount of unsharp masking that may be applied is often the grain in a picture. The sharpness enhancement works on the picture data, and that data contains the grain of the original.

In the case of the modern digital scanner, the procedure of unsharp masking is often applied not optically, but by manipulation of the image data as it passes through short term electronic memory. Three streams of image data are processed, in real time, by a very high speed microcomputer devoted to the task of unsharp masking alone. The microcomputer uses a program containing mathematics derived from the optical domain, so that the effect of digital unsharp masking is similar to that achieved with optical systems. Digital unsharp masking has the potential to introduce effects not possible using optical methods, such as unsharp masking applied to particular colours, or different amounts of sharpening in light and dark areas of the original.

The halftone separations

The making of the halftone separations, which is the point of scanning at all, may take place at virtually the same time as the analysis in a rotating drum scanner. In the flat field scanner this is often not possible. The presentation of the colour data to the exposing device is dictated by the configuration of the output device. The collection of the data by the input device may not lend itself to direct output. It is the format and the frequency of the data that is important here. In repro installations the flat field scanner data is often stored in computer memory in order that it can be re-formatted, processed for colour correction, *etc*, and re-timed for presentation to the output device.

The output device may be one of several possible configurations, but all must conform to the necessary specification dictated by the quality of the output separations required.

Mechanical precision is required to a high degree. If the scan lines of the output device do not lay accurately side by side then the picture will have lines through it, called 'banding.' This often happens in the office equipment output devices which are becoming common place. The scan 'raster,' the pattern of lines formed by the scanning beam, must be accurate in its pattern and, equally importantly, must lay out the same pattern repeatedly on successive exposures. If it does not, then colour separations will not be the same size and, therefore, will not register.

Types of output scanner

Rotating drum

The rotating drum scanner (see Figure 5.11) is the most common configuration of output scanner for halftone colour separations. Unexposed graphic arts film is held, typically by vacuum, to the outside of the drum which is rotated. A traversing optics head contains the exposing source and modulator. The exposing source is often a laser, argon-ion (blue-green) and helium-neon (red) being the most popular. The beam is on continuously but is interrupted by the modulator. This is usually a crystal that has the property that its crystal matrix is affected by a voltage applied at right angles to the light path. If the light is polarised, the crystal with voltage applied will refract or reflect the light, so as to prevent its passage. It acts, therefore, as a light switch.

The voltage required depends on the size of the crystal, so the use of a laser beam requires a smaller crystal and so lower voltage. The reaction time of the crystal is also dependent on its size, the smaller the crystal the faster it can react.

In order to provide good halftone dot structures, the output is required to be made at high resolution. This is achieved by splitting the laser beam into a number of separate beams each acting independently. As a group, the beams are working to produce halftone dots of a size dictated by the analyse scanner. Individually though, the beams are under the direction of individual beam computers. These beam computers are programmed to determine the switching times of each beam according to the film position, the required halftone dot size, the screen ruling, and the screen angle. Usually a choice of screen rulings, angles, and dot shapes are provided.

Fig. 5.11 Rotating cylinder output

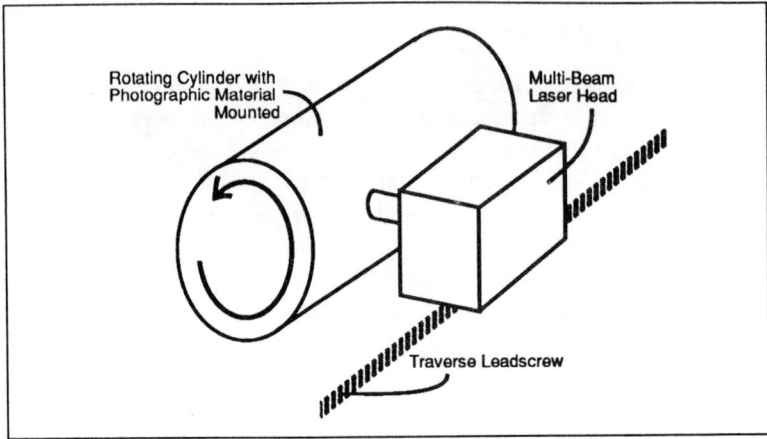

Rotating Cylinder with
Photographic Material
Mounted

Multi-Beam
Laser Head

Traverse Leadscrew

Rotating polygon

The rotating polygon configuration (see Figure 5.12) was developed from the typesetter output devices which they resemble. Such devices are often called 'imagesetters' to distinguish them. They differ from typesetters in that the film transport is greatly enhanced in terms of accuracy

A laser beam is pointed at a rotating polygon, which has a series of accurately ground and polished reflective faces. The reflected laser beam is thus swept over the surface of the film continuously from one side to the other. The film is in continuous motion beneath the sweeping beam, being driven at low speed by a capstan friction drive. A number of beam passes are required to form a line of halftone dots.

Fig. 5.12 Rotating polygon output

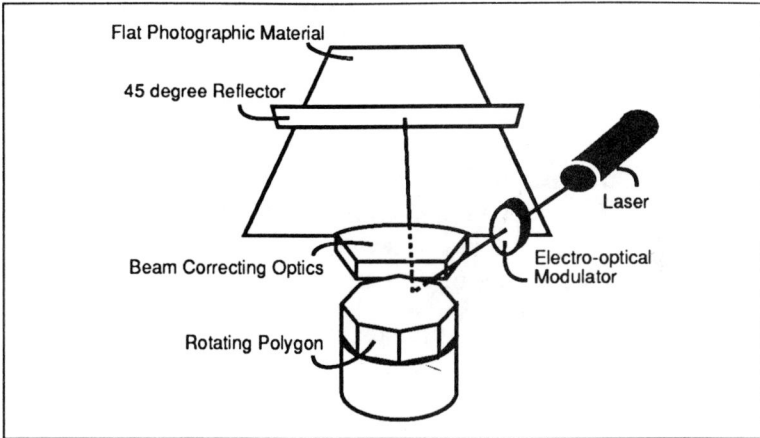

Flat Photographic Material

45 degree Reflector

Laser

Beam Correcting Optics

Electro-optical Modulator

Rotating Polygon

Internal drum

The internal drum scanner with rotating optics is also based on the lathe design of the rotating drum scanner (see Figure 5.13). In this design the photographic film is held on the inside of a static cylinder. A rotating optic head is traversed axially through the cylinder exposing the photo-material in a helical trace. The formation of halftone dots is similar to the rotating polygon design since these devices usually use a single beam for exposure. Fairly high positional precision is assured in this design.

Fig. 5.13 Internal static drum scanner

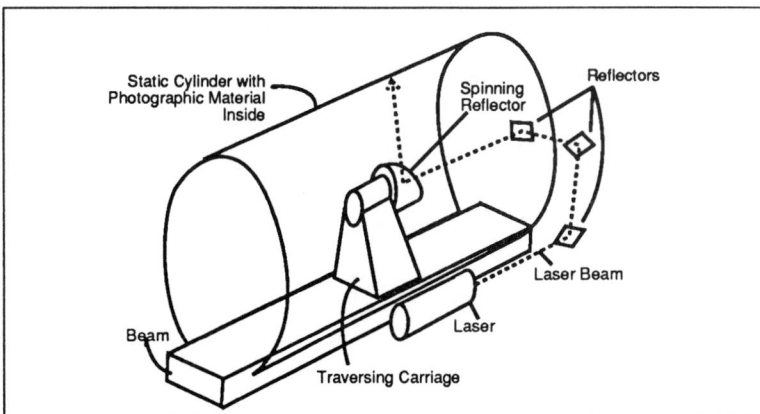

Static Cylinder with Photographic Material Inside

Spinning Reflector

Reflectors

Laser Beam

Laser

Beam

Traversing Carriage

The rotating optics design lends itself to very high mechanical speed operation. The mass of the spinning optic is low and so can be rotated at high speed (12,000rpm is typical).

Resolution

The resolution of a scanner output unit provides an absolute limit to a number of interdependent requirements. By output resolution is meant – the number of discrete points per unit length in two planes on the surface of the output medium.

Most output devices which provide electronic dot generation, are only capable of exposing two different levels, either solid or no exposure. Since the device must be able to produce a solid (opaque) area on the output medium, it follows that each discrete, exposable point must overlap to some extent with its neighbour. Each discrete point is exposed in a particular shape (usually circular), and the resultant matrix of round 'spots' provides the finest representation of any detail in the reproduction (see Figure 5.14). The finest detail in a halftone is the smallest halftone dot that can be produced. For the production of halftones for printing purposes, the ideal output resolution is around 1,000 lines per centimetre (approximately 2,500 lines per inch).

Fig. 5.14 Output resolution – the scan lines and beam switching points form a matrix known as a raster

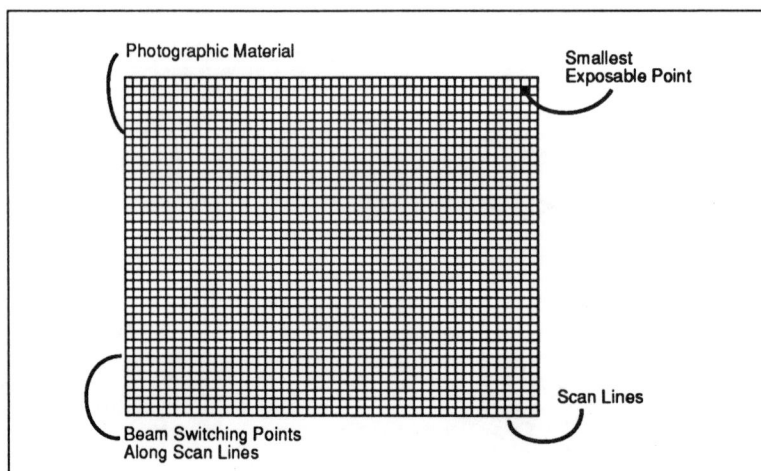

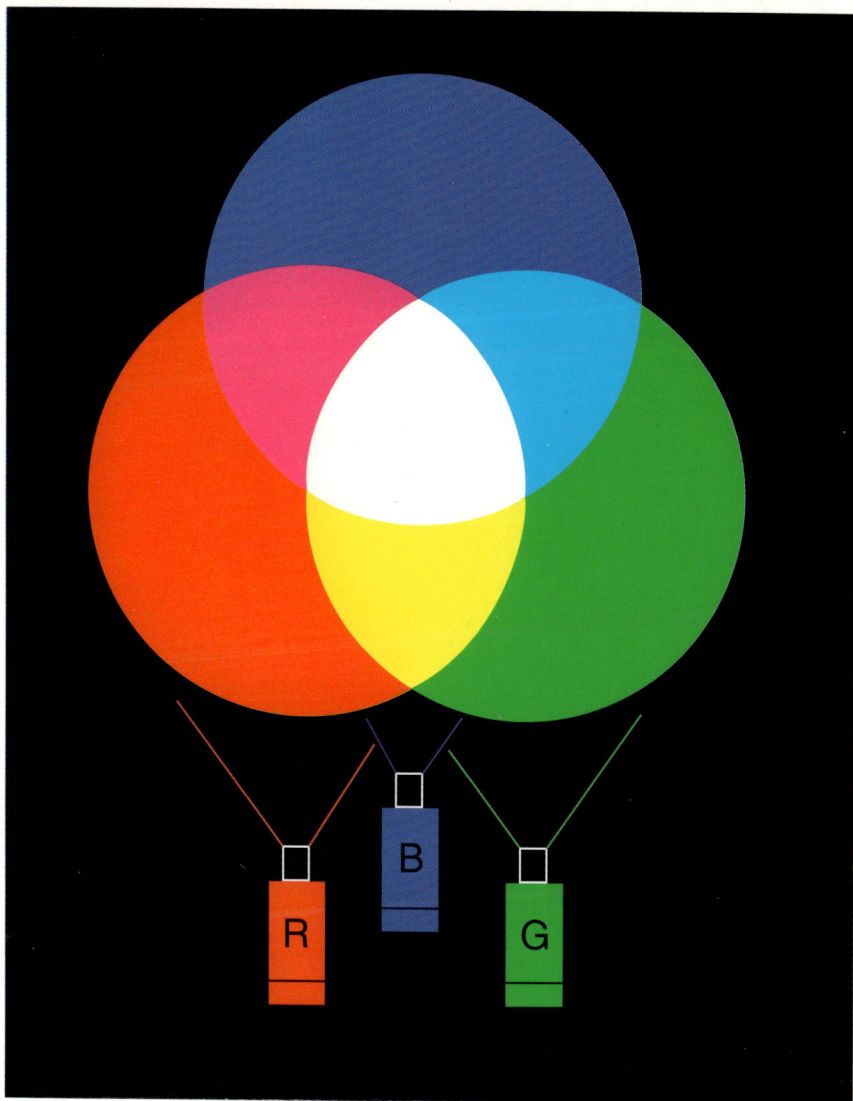

Plate 1 - The additive mixture of colours

Three projectors illuminate a white screen. Each emits one of the primary colours of light - red, green or blue. The beams are arranged to overlap in the way shown. The light beams mix to produce the colours - yellow, magenta and cyan. At the centre all three beams overlap and produce white light.

Plate 2 - The subtractive mixture of colours

Three transparent coloured inks, cyan, magenta and yellow, are overprinted as shown. Three secondary colours are formed, red, green and blue, and one tertiary 'colour' - black. The colours formed depend on the purity of the primary colours and their transparency.

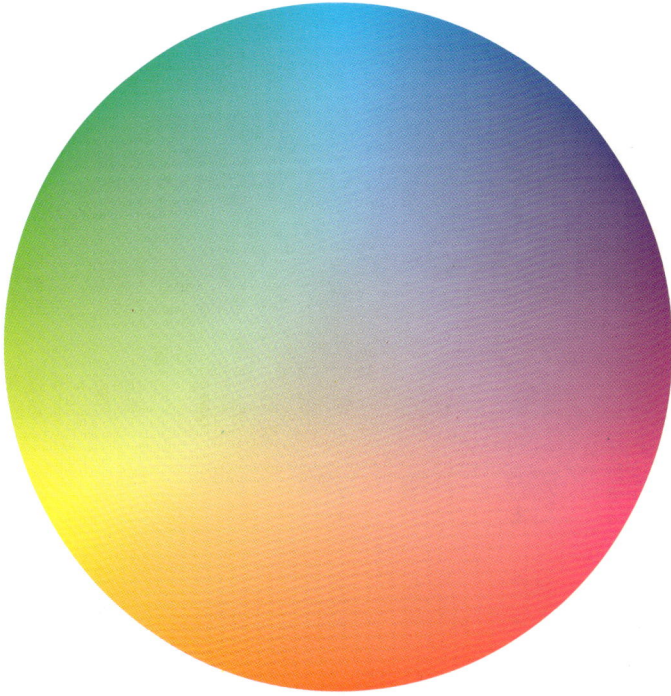

Plate 3 - Cross section through a typical colour space

The space is a sphere with white at the top and black at the bottom (above and below the surface of the page). The colours are arranged round the equator. This diagram is a horizontal section through the equator. Note that grey is in the centre of the section.

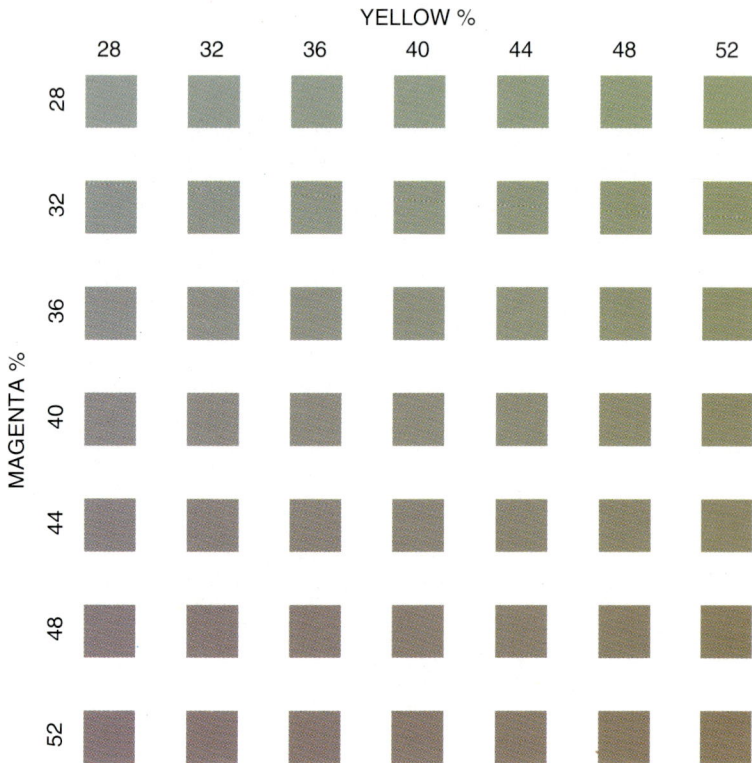

YELLOW %

	28	32	36	40	44	48	52
28							
32							
36							
40							
44							
48							
52							

MAGENTA %

CYAN 50%

Plate 4 - Grey balance print test form

A white is used around the 'grey' samples. The grey sample should be selected and the halftone values for the three colours noted. If the page is viewed too closely, the colour halftone dots may be discernable. Increase your viewing distance if that is so.

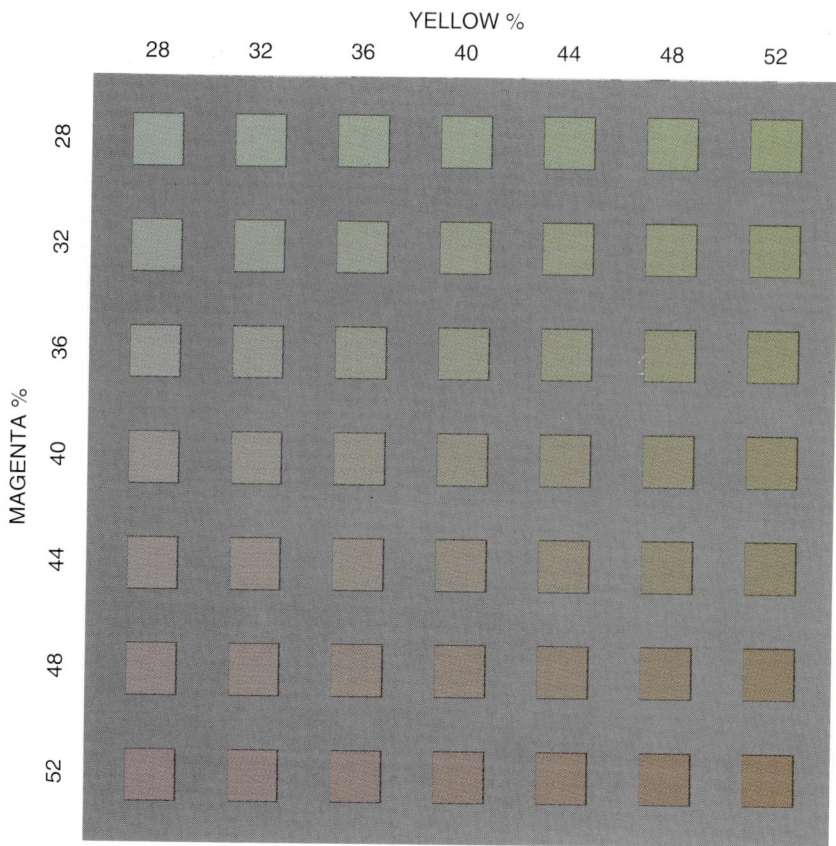

YELLOW %

MAGENTA %

CYAN 50%

Plate 5 - Grey balance print test form

A grey surround, printed in black halftone, is used to aid the grey selection. This has the danger that the selection may be made on the basis of equality of density, rather than equality of 'colour' (grey).

ORIGINAL

| BLUE FILTER | GREEN FILTER | RED FILTER | NO FILTER |

| YELLOW POSITIVE | MAGENTA POSITIVE | CYAN POSITIVE | BLACK POSITIVE |

| YELLOW PRINT | MAGENTA PRINT | CYAN PRINT | BLACK PRINT |

YELLOW **+** MAGENTA **=** YELLOWS MAGENTAS AND REDS

YELLOW **+** CYAN **=** YELLOWS CYANS AND GREENS

MAGENTA **+** CYAN **=** MAGENTAS CYANS AND BLUES

YELLOW
MAGENTA
CYAN
BLACK PRINT
= PRINTED REPRODUCTION

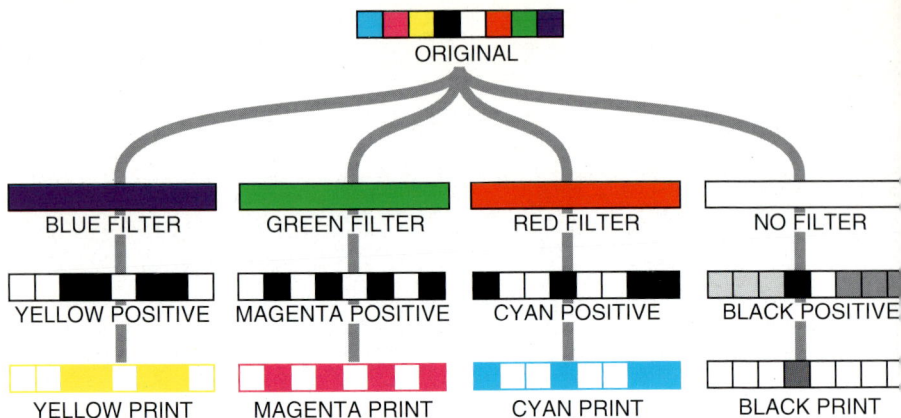

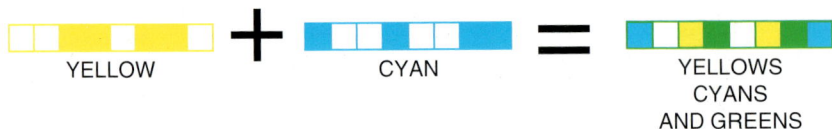

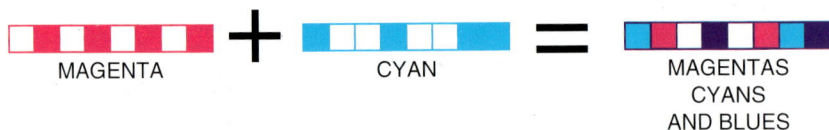

Plate 6 - Simplified colour separation and reproduction

In an ideal system, colour separation by filtration would be perfect. Note the relationships between the colours red and cyan, green and magenta, and blue and yellow. The black separation is adjusted to lose the tone values of the colours in the original.

TARGET	TO CHECK
Solid	Ink density
75%	Dot gain at 75%
50%	Dot gain at 50%
25%	Dot gain at 25%
Slur target	Directional slurring

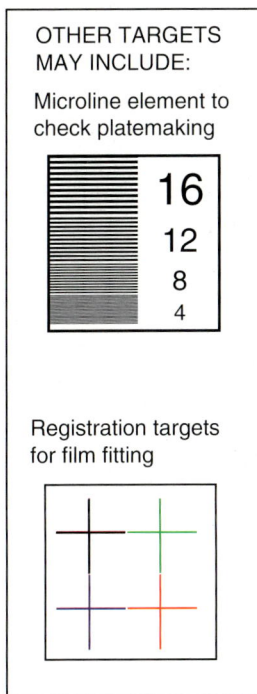

OTHER TARGETS
MAY INCLUDE:

Microline element to
check platemaking

16
12
8
4

Registration targets
for film fitting

Three colour
grey target

Plate 7 - Typical print control targets

The elements are usually arranged in a line, so as to occupy little space. Only first generation films, supplied by the manufacturers, should be used. Print control targets are also called 'colour bars' or 'control strips.'

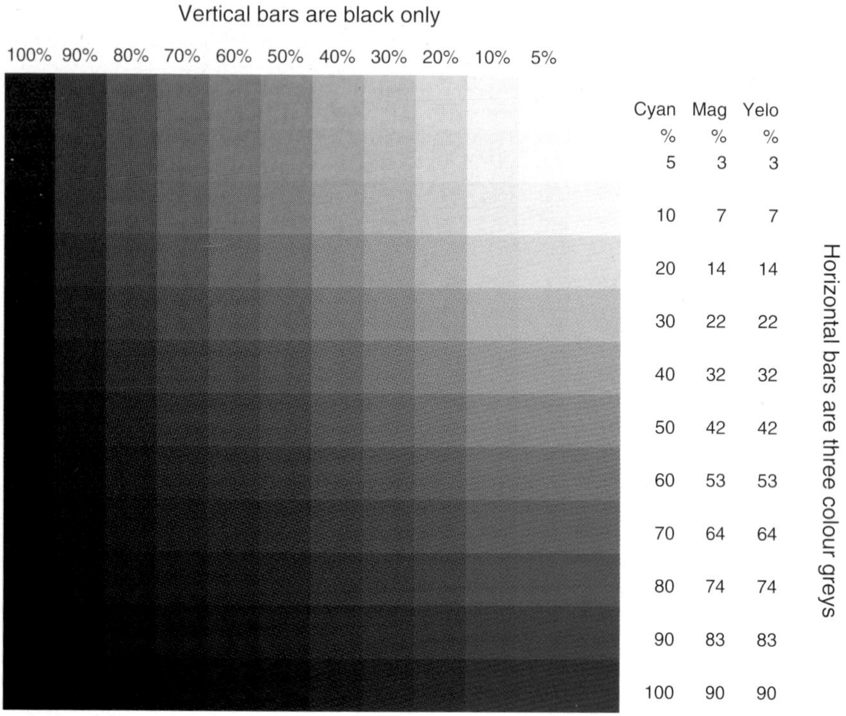

Vertical bars are black only

100% 90% 80% 70% 60% 50% 40% 30% 20% 10% 5%

Cyan %	Mag %	Yelo %
5	3	3
10	7	7
20	14	14
30	22	22
40	32	32
50	42	42
60	53	53
70	64	64
80	74	74
90	83	83
100	90	90

Horizontal bars are three colour greys

Plate 8 - The four colour matrix

The four colour matrix is used to determine the black characteristic referred to in chapter nine. It further defines the four colour print characteristic necessary for the implementation of UCR and the more logical GCR (grey component replacement).

Fig. 5.15 Output resolution grid and halftone screen superimposed (The laser beam is ON in the shaded areas)

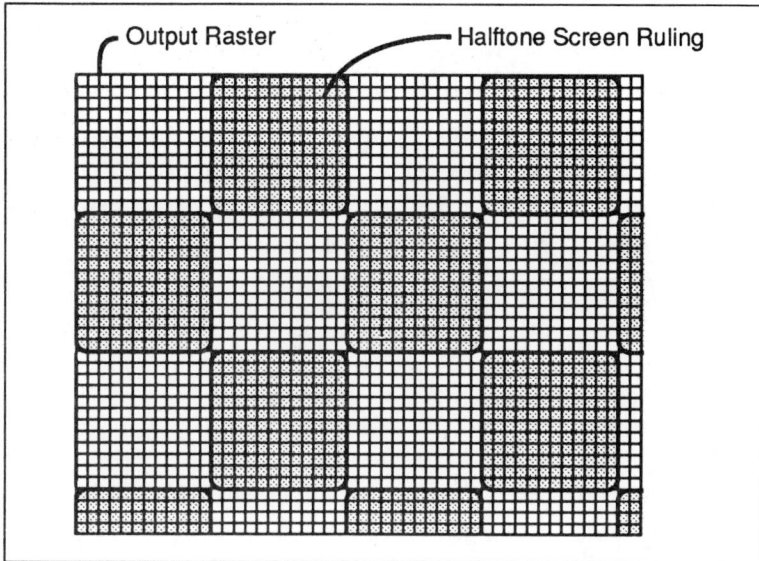

Apart from the ability to expose halftone dots of a well defined shape, the output resolution also limits the number of discrete tone values that may be generated. Screen ruling requirements may vary between 24 lines per centimetre (lpcm), and 70lpcm. Taking an average screen ruling of 40lpcm, then using an output resolution of 1,000lpcm, 626 grey levels can be produced. We can find the maximum number of grey levels in a halftone by the expression:

$$\text{Number of grey levels} = \frac{\text{the square of the output resolution}}{\text{the square of the screen ruling}} + 1$$

Any less than 120 grey levels will be discernible as a series of steps in the reproduction of smooth tone transitions. These are often referred to as contours. Grey levels, in this context, means intermediate density levels of any of the colour separations.

Fig. 5.16 Superimposition of output resolution grid, halftone screen and picture element matrix

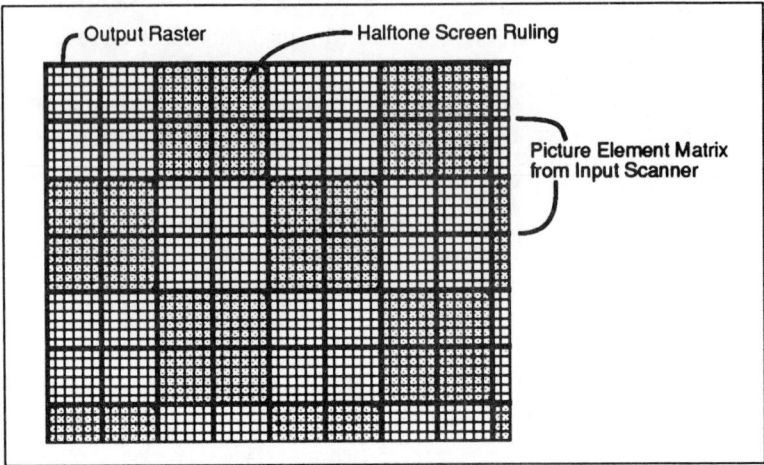

At normal reading distances (about half a metre or 20in) the smallest element discernible as a point is a circle of 0.01mm (0.004in). This provides the lowest usable resolution of pictures intended to be viewed at normal reading distances. If very small pictures are reproduced, the reader is likely to look at them more closely, so reducing the viewing distance. In this case, or where a picture contains a lot of fine detail, higher resolutions may be necessary.

The picture resolution is a consideration only for picture quality. Picture resolution should not be affected, or limited, by screen ruling in a properly designed reproduction system. Although it may be argued that there is little point in producing a high resolution picture that has subsequently to be reproduced by a coarse halftone screen, such an argument ignores the fact that each halftone dot may be formed from a number of picture elements. Each halftone dot in such a case will have a number of sizes – it will be distorted in other words. The distortions of the halftone dot are caused by picture detail (the picture elements) and so will convey a higher resolution than the screen ruling implies.

118

There need be no arithmetic relationship between the picture resolution, the screen ruling and the output resolution in a halftone picture. The 'grids' of each do not have to fit, in other words. In fact if there is no relationship, or if there is an irrational relationship, so much the better. The generation of 'moiré' patterns is a problem of scanning if the precision of the device is not properly controlled. There is less chance of moiré patterns if the relationships of all the constructs are inharmonic.

An ideal scanner for the high quality reproduction of colour pictures will possess the minimum specification shown in Table 5.1.

Table 5.1

Minimum specification for a colour scanner	
Picture resolution	120lpcm
Output (plotting) resolution	1000lpcm
Screen ruling – variable	12-80lpcm
Number of grey levels	>256 (8 bits)
Dot shapes – variable	round square elliptical

Chapter 6 *Image assembly for colour reproduction*

During the printing process we are dealing normally with a number of pages together, as in book and magazine work, or multiple images such as labels and cartons. The individual elements or pages have to be combined together on the image carrier so as to print in the correct position on the substrate and in register with other colours where necessary.

In letterpress printing from hot metal type and engraved picture blocks, the image carrier was a number of individual elements which were locked-up or imposed in a metal frame called a chase. Spacing material called furniture was used to position the elements and quoins (a kind of expanding wedge-like arrangement) locked the type, blocks and furniture in the chase. The resulting imposition of type, blocks, furniture, *etc*, was referred to as a forme. The forme could be used on platen and flat bed letterpress machines but not on rotary machines, which needed curved duplicates or wrap-round plates.

Flexography often uses a number of individual image carriers, in the form of rubber duplicate plates. These are positioned and stuck to the plate cylinder with double-sided adhesive.

Some rotary letterpress machines producing text books also use rubber or plastics plates. They are not stuck directly on the plate cylinder as in flexography but to a thin plastic sheet which is then attached to the plate cylinder.

In those processes where the printing image carrier is formed by photographic means, which includes modern letterpress and flexography, the construction of pages is carried out on film. This operation is referred to as planning, assembly, stripping or montage. There are a number of different methods and materials used to suit the nature and complexity of the assembly.

Goldenrod assembly

The Goldenrod method is most suited for negative film assemblies where a high degree of accuracy is not required, such as in the imposition of text pages in book work or multiple assembly of headings.

The Goldenrod paper is tinted yellow-orange to prevent the transmission of actinic light during exposure to the image carrier. For small litho presses Goldenrod paper can be obtained on to which has been printed information such as grip margins, maximum image area, guide grids, and common sheet sizes.

If a preprinted sheet is not used the layout is drawn directly on to a plain sheet of Goldenrod using the same method as used to prepare a normal layout. The negatives are positioned and fixed with adhesive tape to the layout, emulsion side uppermost. The sheet is then turned over and placed on a light table. Windows are cut in the Goldenrod sheet according to the image areas of the film negative, taking care not to cut through the film.

If a higher degree of dimensional stability is required or if large areas have to be removed from the Goldenrod sheet, making the assembly difficult to handle and the Goldenrod likely to tear, a slightly different technique may be used. The layout is drawn up on the Goldenrod in the same way and a clear plastic sheet is attached on top of the layout. The film negatives are positioned and fixed to the plastic sheet. Windows are then cut in the Goldenrod leaving the plastic supporting both the film negatives, and the Goldenrod mask.

Assembly on peelable membrane

The peelable membrane material is marketed under a number of trade names, including Rubylith, Amberlith and Automask. It is a clear plastic sheet on to which a thin red or amber membrane is laminated. The membrane can be cut and peeled from the base plastic. It is suitable as a mask and carrier sheet for negative assemblies, but is also used to produce burn-out or other masks for positive assemblies.

When used for negative assembly a layout sheet is drawn firstly on cartridge paper or similar material. The peelable membrane material is placed on to the light table, clear plastic uppermost, with the layout underneath. The negative film elements, emulsion uppermost, are aligned to the lines on the layout sheet and fixed to the plastic sheet with adhesive tape.

When the assembly is complete it is turned over, and in areas of the image the membrane is cut and peeled away. Using this method for each assembly would prove very costly, as the membrane material is expensive. If a large number of identical layouts are to be assembled, *eg* book work, it is more economic to assemble each imposition of page negatives on separate plastic sheets. The peelable membrane is then placed on top of each assembly, during plate exposure, to act as a mask.

Positive film assembly

Positive film elements are easier to position and greater accuracy is therefore obtained. For this reason the majority of colour work and intricate assemblies will be carried out in positive form.

For monochrome or simpler register work the positive film elements are assembled on to a clear plastic sheet, called a foil or flat. Foils are made from a dimensionally stable plastic such as polyester. The correct positioning is determined by the drawn layout sheet placed under the plastic foil. Where two or more colours are to be printed, the position of the film elements for the second and subsequent assemblies is determined by registering them, by visual alignment, to the first assembly.

Film assembly may be employed to provide finished pages in a book or magazine. The pages are often 'finalised' by contacting on to film, which process is used to incorporate all the elements of the page into one final film. Such final film pages are mounted on to the imposition foil to achieve correct page layout, and orientation, for folding.

Blue and red key assemblies

Obtaining perfect register when visually aligning one film element to another can be difficult and tedious, particularly in
process colour halftone assembly. Improved accuracy can be
achieved and the task made less tedious by the use of a blue or a
red key. The material used to produce the key is a polyester sheet
with a light sensitive coating. When exposed in contact with a
film assembly and developed, a sharply defined image is produced in negative, positive, blue or red depending on the material being used.

Using this method we first select the separation, from each
process colour set, which has the greatest detail. These key
separations are fixed to a clear plastic foil using the drawn layout
to obtain the correct position. The assembly is then exposed in
contact with the key material to produce a colour key.

The blue or red key is then used in place of the drawn layout,
separations being individually registered to the colour key. The
blue key image will transmit actinic light, and the separations
can, therefore, be mounted directly on the key sheet itself and
then exposed to the plate. When using the red key a clear plastic
foil is overlaid on which the separations are mounted.

Film assembly by pin register

The aim of pin register is twofold, firstly to improve accuracy of
layout, and secondly to reduce time spent on make-ready on the
press. By relying on mechanical means for registration, it is
intended to obtain a consistent accuracy and not to rely on human
intervention. The responsibility for initial registration is with the
planner but once registration is achieved, it can be regained by
the use of register pins.

The layout sheet and the carrier or mounting foils on to which the
separations of each colour are to be assembled are punched on a
plate/foil punch (see Figure 6.1) and the foils are then located on
master location pins set into the light table of a registrator. The
registrator is a large illuminated table over which a gantry runs
carrying a rotating hollow drill or punch (see Figure 6.2). The

hollow drill is connected to a vacuum pump which withdraws the perforated foil material after drilling. With large foils the tail edge may be stuck down with tape to the table to prevent movement.

Fig. 6.1 Example plate/foil punch hole pattern

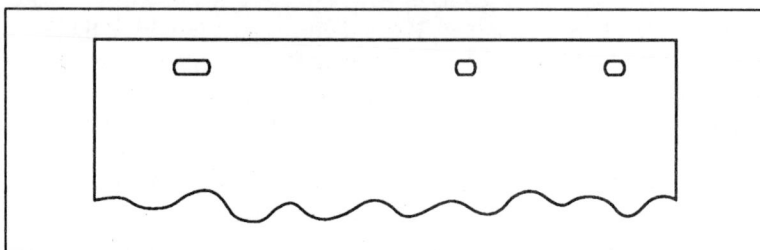

The punched separations are then assembled on to pins and positioned manually and visually to the layout sheet below. A weighty film location bar is placed on top of each set of separations in order to keep the films in close contact and the pins vertical. When location with the layout is obtained the sets are stuck to the top carrier foil with adhesive tape to hold them in position.

Fig. 6.2 Foils and films on register pins are drilled to combine the two

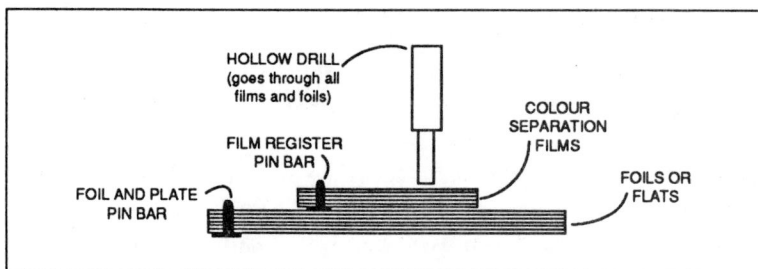

Holes (usually two) are then drilled in the edges of the sets and also through all the carrier foils. It then remains for the separations of each colour to be assembled on their respective foil and stuck down.

The plate/foil punch is used to punch the plates at exposure time. The foils, mounted with their film images, are then exposed to the plate with pins inserted. The pins pass through both the foil and plate, ensuring that the images of each colour are exposed in the same relative position on each plate. Any burnouts or masks are also punched to ensure register.

After exposure and processing the plate is mounted on the press plate cylinder. The cylinder is also fitted with register pins, so once again the plates mount in the same relative position on each cylinder. This helps to ensure the rapid attainment of register on the press during make-ready time. For web presses, the plates are required to be accurately bent, in order to locate the plate lock-up system. In this case the register pins are housed in the plate bending jig.

Electronic page composition

From the foregoing it can be appreciated that there is no set formula for the way a job is planned or assembled. What is equally clear is that a lot of time and film are expended in this part of the process. The attention of one operator is required from the start to the finish of the process: It is difficult for operators to take over partly completed assembly jobs. They have to work out the method employed by the first operator before they can continue the process.

For reasons of time saving, materials saving, and continuity, to say nothing of accuracy, the film assembly part of the process has moved into the electronic domain. Electronic page composition (EPC) began in the late 1970s and was based on the then popular minicomputers. Such computers were built from discrete components, or basic logic chips, and used large magnetic disk drives for storage. They quickly developed to the point that more than one computer was required and the systems could be purchased with three or four computer 'tracks'. The images in use are typically displayed on a high resolution display monitor where attention is paid to the appearance of the image on the screen. They are often calibrated to preserve that appearance. Much of the methodology employed in EPC systems came from the image enhancement work carried out for space research.

An EPC system consists of an input or acquisition device; a computer or computers, of high power and with a great deal of memory; and an output device. It is fortuitous that the input and output scanners in use in the graphic arts are ideal for use as input and output devices. In action the EPC system works as follows:

> The image is input on the scanner in exactly the same way as usual. The scanner operator sets up the input scanner to produce the required set of four colour half-tone separations.

> Instead of the scanner signal being processed and used to drive the output scanner and so produce the colour separations, it is stored on a magnetic hard disk – either a 'Winchester' drive or a 300megabyte removable disk drive.

> The image data is scaled to fit the video monitor which can display, in pictorial terms, the contents of the memory.

> The computer is used to display an image of the page into which are placed the scanned images in their correct positions.

> Other elements in the page, such as tint blocks and borders are constructed on the computer, also in the correct position on the 'page'.

> The computer runs through a routine where the magnetic disks are read in the correct order to produce a new magnetic disk containing the data of the page in scan order – a process referred to as assembly or execute.

> The output disk drive is read and the data sent to the output device which exposes film images of the complete pages.

The scanner time is extended by this method, that is, the scanner scans in only, then the page is made up, then the scanner scans out only, actually taking three times longer than the direct scan to halftone film separations. If the sum of the computer and output times is less than the equivalent film planning time, then the electronic process is quicker. Bear in mind that there is often some extra overhead in the electronic process – that of file transfer. This is non-productive time for the image to be transferred between disk drives, displayed on the screen and assembled.

The accuracy of electronic page make-up is near absolute, *ie* images are placed within one picture element.

Electronic page composition has developed with the power of the available computers. Memory prices have reduced significantly in the course of time, and the speed of the computers has increased. This means that the designers of these systems can be more generous with memory capacities, allowing the processing of image data in electronic memory rather than magnetic memory. This speeds up the processing and access times.

Developments to the systems have enabled the introduction of very advanced image manipulations such as picture element copying or 'cloning'. Cloning allows for the overwriting of one part of an image with data from another part to effect invisible retouching, for the removal of flaws in the original. Most picture information contains texture, and the cloned data retains the texture of the original image, When retouching a skin tone, for instance, the retouching 'medium' can be seen as a colorant which has the texture included. Typically this is achieved by displaying two cursors on the screen, one of which indicates the source of image data, and the other the destination of that data. If a scratch on the original is then 'painted' with the destination cursor, the area is replaced with image data from the source cursor position. Both cursors normally move together, in parallel, so that the painted image data is not repetitive. Obviously some care is needed in making the choice of the area of the source data as it must have a very similar appearance to the destination area.

EPC systems often include functions which imitate airbrushes and other traditional retouching tools so that extensive creativity is also available. It is now possible to assemble images and other elements together which would be nearly impossible, and anyway very expensive, by traditional means.

The more complex the requirements of an EPC, the more power, speed and memory capacity is required in the computer. Current office computers have more of all these features than the early EPC systems. It is then easy to see how desktop publishing systems can emulate many of the facilities of EPC.

Imposition

One of the prime requirements of any printed product is that image areas should be located consistently in positions required by the customer. The degree of accuracy involved will vary according to the product, from fairly generous tolerances in single colour work, to that where differences of 0.05-0.1mm (a half to one halftone dot typically) will become apparent visually or interfere with other processes or end product performance.

This requirement is made more stringent when multi-colour work is involved. Here the printer is not only concerned with ensuring the correct position of the original image areas, but also that subsequent images superimposed on the same printed sheet will register with each other. Note that the degree of accuracy required depends very much on the kind of detail in the colour sets, so that, for a large number of jobs, there may be three location conditions to satisfy:

The image is in the correct position relative to the material on which it is printed.

Different images on the same sheet are correctly placed relative to each other.

That images are sequentially superimposed accurately one upon the other, eg in process colour work.

The planning or imposition of the film material to meet these location conditions will normally be carried out using one, or a combination, of the following methods:

Visual register of images to the layout and of one colour to another.

Visual register using colour keys.

Step and repeat machines.

Each of these methods is more fully described below.

Visual registration of film images

This method simply entails the visual positioning and fixing of film images, of one colour (the key), to a clear plastic sheet (foil), the correct position for each image being indicated by a layout sheet placed under the plastic foil.

The film separations of the remaining colours are then positioned and fixed into their appropriate foil, using the imposed foil of the key colour to indicate their position.

This method can provide accurately imposed film images cheaply and quickly, but is very dependent on the skill of the planning personnel. The speed and accuracy of imposition is affected adversely if negative film images are used, positive images being easier to position visually. It is also a great advantage to have accurate register marks on each film image, which may not always be possible.

Visual register using colour keys

Using this method the key colour separations are, as with the previous method, fixed to a clear plastic foil using a drawn layout to provide correct positioning of each image.

The imposed foil is then exposed in contact with a light sensitive coating on a polyester sheet. The exposed sheet is then developed to produce a sharply defined image in negative, positive, blue or red depending on the material being used. This image is then used as the master for visual positioning of the remaining film separations. The contrast between the colour key image and the film image facilitates more accurate and less tedious registration of the film images. The contrast is particularly advantageous when there are no register marks, and detail within the image itself has to be used.

Visual registration should not be considered as a poor compromise in planning. It is sometimes the case that a small transparency, given a fair degree of enlargement, will not fit (register) itself. The emulsion of the transparency typically contains four or five separate coatings of gelatine. The yellow, magenta and cyan dye images occupy different layers in the emulsion. If, for reasons of speed or poor practice, the transparency is not properly acclimatised at the time it is separated, it may change size during separation. In this case the separated images do not register. Visual registration may allow for the mis-register to be 'shared' across the image, or to be placed in an area of the image where mis-registration is not easily seen.

Step and repeat machines

For certain types of work such as magazines, books, catalogues, packages and labels, where a number of different full page separation films are to be positioned and copied on the same plate, or the same separation film repeated a number of times, the step and repeat machine can be employed usefully. This method eliminates the need for planning on to large mounting foils thereby reducing costs, especially in materials.

At its simplest the step and repeat machine is a set of crossed gantries which enable a film carrier to traverse an unexposed plate in two dimensions. At preset positions the carrier will contact and expose the image on the film to the plate surface by means of a lamp mounted on the carrier. In this way the plate is successively exposed with a number of identical images in position on the plate's surface. Step and repeat machines may be motorised and computer controlled so that the operator has only to set the step and repeat distances, load the film carrier, and set the machine going.

More recently the same principal has been employed to achieve impositions even where the image is not duplicated. In this method the machine, which is similar to a step and repeat machine, has an additional stacker to one side. The page films, punch registered, are mounted on the stack in the correct order, usually on a jig frame. The machine then picks up the first image and traverses to the correct position on the plate where the film is exposed to the plate. The gantry returns the first film to the stack, picks up the second and the process is repeated for each page. This method saves the time involved in imposition and allows for rapid changes in the folding procedure.

Platemaking

A number of alternatives are available in the making of the printing surface, although the printing process dictates, of course, the exact nature of it. In the gravure process it may be that the cylinder is made by electronic means, being engraved by a stylus under the control of the stored image data. The cylinder may also be made by photochemical means.

A flexographic process demands the use of rubber or, more likely, a special polymerised resin, whilst litho requires a flat hydrophilic surface able to carry an oleophilic image. Both flexo and litho are likely to have their plates made by largely manual means but both processes are served by laser platemaking equipment and so may have their printing surfaces produced by machines, again under the control of image data.

The most representative platemaking process, which is litho platemaking, is also the most common, so this chapter describes that process.

The positive or negative films are the input to this process. They may be supplied as final, one piece films, or as plastic foils (sometimes called flats) upon which are mounted the colour separations and other images in their final printing positions.

The printing plates are mostly supplied as presensitised aluminium sheets of the correct size to fit the printing machine. By presensitised is meant that the aluminium sheets are coated with a light sensitive (usually ultra-violet sensitive) coating. The coating will form the image layer of the plate and is therefore oleophilic after processing. The coating is usually a resin-based polymer which may be positive or negative working.

The process consists of accurately positioning the foil on the surface of the plate, placing the combination into a vacuum frame, and exposing the plate to powerful ultra-violet radiation. In this way the plate coating is affected by the light either by hardening it, in a negative plate, or breaking down its molecular structure in the case of a positive plate. Chemical processing is employed to clear the unneeded coating, and leaves the image areas on the surface of the plate ready to accept ink on the press. Certain of the coatings types used may be hardened further by thermal means to extend the press life of the plate.

Platemaking control

The increasing use of presensitised plates as opposed to deep etch for colour printing has simplified the platemaking process, and placed less of a demand on the skill and craft of the platemaker. In particular the burden of responsibility for consistency of raw materials, once that of the platemaker, is now primarily on the shoulders of the plate manufacturers. Nevertheless, accurate control of platemaking at the printing down and processing stages is still essentially in the hands of the platemaker.

Before discussing in detail the areas which need control and how this is best achieved, it is helpful firstly to define what is required in general terms for a lithographic plate.

A photolithographic plate should be processed to give:

> *The maximum plate life possible.*

> *Maximum difference in wetting, between image and non-image areas (non-image = hydrophilic, image = oleophilic)*

> *Consistent reproduction of the image on film, negative or positive.*

These are primarily dependent on correct exposure and processing of the plate.

The correct exposure times are determined using a continuous tone step wedge (Stouffer wedge) and maintained with the aid of a light integrating unit. A light integrating unit has the appearance of, and operates as, an exposure timer. If the light output of the lamp changes for any reason, the integrating unit extends the exposure time to compensate. Consistent exposures are made possible by the use of the light integrator. The improved light output from modern exposure lamps and the greater sensitivity of plate coatings has resulted in much shorter exposure times than were required in the past. This has increased the importance of accurate exposure, as slight variations can have a more significant effect on the overall exposure.

The Stouffer wedge consists of a piece of film on which there are 21 numbered steps of continuous tone. Each step increases in density by 0.15, step 1 having the lowest density. When exposed in contact with the litho plate, each step allows a different amount of light to reach the plate coating. When the plate is then developed some steps will be solid, others will be grey or not visible at all. The plate manufacturers will normally, in their instructions, recommend a step to which the plate should be developed. For example, negative working plates solid step 6; positive working plates clear step 2. It is important that the sensitivity guide is placed under all film layers at the outside edge of the plate.

Correct processing is achieved by good housekeeping and close adherence to the manufacturers' instructions for development and desensitisation, *etc*. The use of automatic plate processors, if maintained in good order, is an aid to achieving this consistently.

Consistent reproduction on the plate, of the positive or negative image, is the most difficult of the three factors to achieve but is most important, if faithful and consistent colour reproduction is to be achieved.

Factors which may influence the accuracy of image transfer to the plate include:

> *Quality of film positive or negative.* All negative or positives used for platemaking should have been produced by contacting to give second generation hard dot film image (finals).

> *Resolution capability of the plate.* Plates from different manufacturers are likely to have different resolution capability, particularly where there is a marked differ-

ence in the grain, or if different types or thickness of image coating are used.

Contact in the vacuum frame. Poor contact between the film and plate in the vacuum frame will allow light to undercut the film elements, causing image sharpening on positive working, and thickening on negative working plates.

Exposure. Over-exposure of a positive working plate may cause image sharpening and on a negative image thickening or gain. The extent to which this happens will be influenced by the type of plate and contact in the vacuum frame.

Use of diffusion film during exposure. If a diffusion film is used to reduce the effect of film edges, on positive working plates, under cutting is likely to be more pronounced. The extent will again be influenced by the type of plate and vacuum contact.

The accuracy of image transfer is best judged and controlled by the use of one of the many control elements designed for this purpose.

Line elements are used in the FOGRA PMS and Gretag-UGRA PCW control elements. Both contain fine lines of varying thickness adjacent to spaces of the same thickness. In the PCW wedge these lines take the form of circles. If differences in image transfer occur between plates it is shown by the fact that different line thicknesses will be reproduced on the plates. Both suppliers of these elements provide tables which enable the user to assess the dot gain or loss occurring with a specific line reproduction. Pira's experience suggests that these tables are somewhat questionable but, nevertheless, as a comparative tool the control elements are very valuable.

A number of control elements include very fine halftone dots for the assessment. The Brunner colour control bar, for example, contains six different dot sizes with areas of 0.5, 1, 2, 3, 4 and 5 percent respectively. If differences in image transfer occur then the plates will reveal a different step being held.

Some elements surround the fine dots with those of a far coarser screen ruling. Since these dots are far less sensitive to variation they act as a reference point. The same principle is used for visual assessment of dot gain on the press and we shall return to it later.

The Baker control strip is for use in both the plate room and the press room. The signal strip comes in a continuous length, ready for printing down. It consists of three sections which are very sensitive to changes in image quality. The two variables we can detect in the plate room with this device are:

> *Contact in the vacuum frame.* Contact can be detected in the print-slur section by moving a light in a low angle arc over the frame – if a pattern appears, the contact is poor.

> *Variations in dot size between finished plate and signal strip.* Dot variation can be detected in the plate-guide section by examining the quality of this section on the finished plate. If no pattern appears, the dots on the plate are the same weight as those on the signal strip. If the narrower bars are light, it means the dots on the plate are sharper than those on the film. If the narrower bars are dark, the dots on the plate are heavier than on the film.

The GATF dot gain scale is similar to the plate-guide section of the Baker signal strip, but here a number between 0 and 9 indicates the degree of dot gain or dot sharpening. The advantage claimed is that variation of dot size can be given a numerical figure.

Control elements for platemaking are particularly important if 'soft' dot negatives or positives are used. Relatively small exposure differences can give rise to significant variations in image transfer in this instance. Rigorous control is then particularly important and elements containing images of the type described above are absolutely essential.

Whilst the use of control elements for the specific purpose of platemaking control are valuable when setting up a platemaking procedure, their use as a routine check in production conditions is inconvenient. Many of the control targets for use in proof and print control contain platemaking check elements. The use of print control targets therefore enables the checking of the platemaking procedure on a continuing basis. Print and proof control targets are described in the next chapter (see also Plate 7).

Chapter 7 *Colour proofing systems*

Introduction

The finished product of the reproduction processes can be either a set of colour separation halftone films or a set of colour printing plates, neither of which has the appearance of the printed job. Ideally the plates should be installed on the press and run to produce a satisfactory job. Since press time is expensive, and press performance a possible variable, it is normally necessary to prove that the films are capable of producing the required result. The colour separation halftone films or plates are 'proofed', to produce a coloured rendering of the job, for approval by various agencies. The reason for producing a proof often dictates the essential parameters of the proof. Ideally, of course, all proofs should be identical with the finished result in all respects. This is not possible for a variety of reasons.

When the job is in the early stages of planning, the text may not have been finalised, the pictures may not be complete in number, or the accompanying graphics may not have been decided. Designers have in their minds a vision of the completed job which they are unable to share with other parties. Once the job is under-way, there is a delay until all the pictures are scanned, processed and placed in page position. At this stage, typically, the separation films are ready for production, and if they are produced, they may be proofed by an off-press proofing method, of which there are many. As an alternative, plates can be proofed on a specially constructed proof press. If customer approval is not given, the repro process is re-entered for the purposes of correcting the job by varying the films. Several iterations may be necessary before the job proceeds to be printed.

The proof press is generally a flat bed machine having two platens. On the first of the platens, the plate is mounted and on the second, individual sheets of paper are clamped in a position defined by metal tabs called lays. Across the top of the two platens passes a rolling gantry, containing a cylinder which carries the litho blanket and inking/damping rollers. As the gantry passes over the platens, the plate is damped and inked. The blanket then rolls over the plate, transferring the image to the blanket. The blanket next drops into contact with the paper on the second platen, in a continuous rolling action, so transferring the ink image to the paper.

The job may be interrupted for the production of 'scatter proofs', which are produced for the purpose of checking colour accuracy, before the colour sets are invested with further work. Photo-polymer pre-press proofing systems have proved popular in this application. Such systems save the costs of providing facilities for, or of buying, ink-on-paper proofs and provide the means of in-house evaluation in the repro houses.

Photopolymer-based systems are confronted, technically, with different problems to those of printing. Their colorants have to work within the system, have to be safe to handle, and have to be cost effective in relation to the ink-on-paper proofing system they seek to replace. On top of their own problems, they have also to synthesise the problems of the printing process – a formidable set of criteria. In fact, an impossible set of criteria. It is a great credit to the manufacturers of such systems that they work as well as they do. From time to time the manufacturers are confronted with problems such as matching different printing processes (*eg* gravure and flexo) or matching different printing characteristics (*eg* newsprint in litho or low gain litho presses). The manufacturers have a good record in approximating all these requirements.

Proofs may be required at many stages in the reproduction chain, from early 'visuals' to final customer approval proofs. They may be described by their purpose:

> *Visual proof – the designer's way of getting their ideas across to the publisher or customer.*
>
> *Typographical proof – which may be a bromide or Ozalid monochrome proof for the purpose of reading the text.*

Scatter proof – for the internal checking of the accuracy of colour work before proceeding to film planning stages.

Target proof – for guidance purposes during press make-ready.

Contract proof – the customer approval proof which is often signed by the customer.

Scatter, target and contract proofs are the proofs which must convey an accurate impression of the finished job, and therefore be a good colour match with the resultant press sheets. The proofing system must be able to simulate the visual appearance of the printing condition.

The role of a colour proof is often seen quite differently by those involved in various parts of the industry. To those producing halftone colour separation films, it is used as a means of checking that a satisfactory set of colour separations has been produced, then, once it is certain that this is so, it is the means of demonstrating the excellence of those films to the customer.

To the customer, it is a specimen print, showing that the separations have been made in such a way as to give a satisfactory printed reproduction of the original. They expect that the proof will be typical of the job which is to be printed.

The printer uses the proof as a guide when setting up the press, with the intention of matching its appearance as closely as possible. Difficulties arise when the printer discovers that the proof has been produced in such a manner that it cannot be matched. It may have been printed on a slow running, single colour, flat bed proofing press; possibly from a different set of plates; and in some cases using different inks on a different substrate. Such conditions may produce a proof which cannot be matched realistically on a conventional production press. It is increasingly possible that the proof may not have been produced by ink on paper, but by using one of the photomechanical off-press proofing systems. The difficulties in this situation may be even worse, depending on the pigments or dyes used by the system and the substrate on which it is produced.

The only satisfactory method of resolving these problems is to stabilise the press run and determine its characteristics and then endeavour to match these characteristics on the proof. This will ensure that proofs may be matched, without difficulty, on the press. We will discuss the factors to be considered when trying to simulate press conditions at the proofing stage and suggest possible methods of ensuring that the proof has similar characteristics to the press. Both conventional and off-press proofs will be considered together with the implications of viewing conditions.

Specifying the press characteristics

The significant properties of the production print run which require careful specification are probably self-evident to many experienced printers. However, the importance of carefully establishing such a specification cannot be stressed too strongly since its use both by camera or scanner operator and proofer is fundamental to successful printing. A number of the factors which are totally dependent on the printing condition must be carefully specified, preferably using statistical techniques, and properly communicated to both proofer and camera operator. The printing condition is influenced by a combination of the following variables:

Press.

Paper.

Ink.

Blanket.

Test forms, comprising an assortment of elements such as step wedges, multi-colour overprints and slur doubling targets, are commonly used to establish the press characteristics. These are run on the press under test using the various ink/paper combinations in common use and the information gathered is used as the press characteristic. It is possible to incorporate small test elements in all jobs passing through the plant for a period of say, six months or more, and undertake a statistical analysis of random sheets from each run. Most of the information required can be gathered adequately using colour control strips which are available. Such methods will allow for variations which occur over extended periods, such as changes in temperature and relative humidity – where press rooms are not fully controlled – aging of blankets, and minor variations in blanket packing to obtain register. It is, of course, important that such factors are recorded in order to attempt to establish their significance and, if found to give rise to large variations, that steps are taken to remedy them.

Probably the only important variable which cannot be readily determined by most printers using test elements is that of grey balance. Ideally this requires a special test form such as that marketed by the Rochester Institute of Technology. It is necessary, therefore, that this be included at least once on each ink/paper combination, and preferably more often. A complete description of the procedures for making a press specification is in Chapter 9.

Matching the proof to the press characteristics

Conventional proofing
There are many factors which can give rise to differences between proofs and press sheets. In order to reduce them to a minimum a considerable degree of standardisation is required both at proof and production presses and, once the standards are established, they must be properly maintained.

Before producing a proof, which is to synthesise the production job to be printed later, there are five main factors which must be established:

> *Printing substrate.*
>
> *Colour of the ink to be used.*
>
> *Solid printing ink density.*
>
> *Trapping characteristics.*
>
> *Tone reproduction.*

In practice these are all interrelated as we shall discuss and cannot really be considered in isolation; however a few remarks on each are pertinent.

Substrate
The need to proof on the substrate on which the job is to be printed is well established but needs clarification. In most cases the substrate acts as a reflector for the light falling on it and hence differences in paper colour will cause consequent differences in ink colour. The unprinted paper will generally act as a neutral reference for the eye. The substrate will also seriously affect the tonal gradation of the final reproduction because of its influence on dot gain, a major cause of differences between proofs and production sheets.

Colour of ink
The use of British Standard process inks (BS 4666) has probably reduced the significance of problems arising from differences in ink colour, but problems will arise when non-standard colorants are used. It is of utmost importance in such a situation that sample inks are supplied to the proofer.

Solid ink density
The printed density of the ink is a problem which today is probably not as significant as in years past. A better understanding has arisen between proofers and customers of the limitations within which the pressman must work in order to avoid set-off and smudging. It is less common to see proofs produced at ink weights which are achievable when printing wet-on-dry, a single sheet at a time, but not possible in production printing conditions. It has also been assisted by the trend towards inks of a higher pigment concentration, thereby allowing the pressman to print stronger colours using reasonable ink film thicknesses.

Trapping
Trapping can be responsible for major problems in matching proofs to production printing conditions. It is very rare to find a wet-on-wet press where perfect trapping is obtained for all colours and this will obviously have a significant influence on secondary and tertiary (three-colour) colours. It is the responsibility of the printer to ensure that trapping deficiencies are kept to

an absolute minimum, but they can rarely be completely eliminated. What then is the proofer to do in order to match this? It has been Pira's experience that on many occasions the proofer may partly simulate the effect by changing the proofing ink sequence, and this is worth trying. It is certainly not a complete answer but given the differences between proof and production presses any better solution appears unlikely.

Tone reproduction

It is almost certain that, in the majority of cases where proofs and production prints do not match, differences in tone reproduction (or gradation) are the underlying cause. The problem of dot gain, familiar to all litho pressmen, is something which proofers try to avoid, and in the main they succeed. It is undoubtedly this aspect of matching press results that most needs emphasis.

The causes of dot gain at the press are not easy to isolate. They are generally a combination of a number of factors, including the blanket, paper, pressure, speed of printing, and halftone dot area. However, the most significant factors are the ink rheology and pigment concentration. The relationship between ink film thickness and dot area shows a dramatic increase in dot area as the ink film thickness is increased, and thus in a situation where proofer and printer are using inks of a different concentration the likelihood of differences in tone reproduction is high. However, ink film thickness variation cannot be the only cause of dot gain on the press since we know of situations where both proofer and printer are working from the same ink and where the proofer is still printing a relatively sharp (low gain) dot. Since both are printing to the same solid density, any differences in ink film thickness must be fairly minimal. The likely cause is printing pressure differences, in this case.

Printing pressure suggests another possible cause of dot gain: the influence of subsequent printing units on a wet-on-wet printing press. There are two possibilities which may influence dot size. The flow of the still-wet ink from the previous unit when in the nip of a subsequent unit, and the possibility of a slight double impression giving an increase in effective dot area. The influence of third and perhaps fourth units, on the other hand, can be to decrease the print density in some cases. A microscopic examination shows this to be due to a micro-picking effect within the image with certain ink/paper combinations.

It is almost certain that dot gain is significantly influenced by the rheology of the ink being used and this we believe to be the most likely cause of differences in tone reproduction between proof and production sheet. Even if it is not the primary cause it suggests an important method of reducing the difference. The use of an ink diluted by 'reducers' – non-coloured bulkers – allows the proofer to apply more ink to achieve a given solid density. Such an ink tends to produce a higher dot gain (mid-tone density) than an untreated ink. In this way the proof press characteristics can be tuned to match the printing press characteristics more closely.

This strategy may be combined with print pressure changes on the proof press, which affect the dot gain above the mid-tone to a greater extent, to match dot gain throughout the range of dot sizes.

Platemaking

It has been assumed in the above discussion that both proof and production print are being produced from the same plate, but it is not uncommon, for reasons of economy, to have special proofing plates and not make the printing plates until the proofs are approved. Whilst understandable, this is an unfortunate practice because it can give rise to serious problems. Even if the plates from which the proofs were produced were not intended solely as proofing plates, but became so because of necessary corrections, the problem may still arise. If both are surface plates, and therefore the only significant variable is exposure, then the platemaking process must be properly controlled by the use of suitable exposure targets. Where it becomes particularly difficult is when either proof plates or production plates are of different types.

Off-press proofing

The use of photomechanical methods of proofing is now common, and their use as customer proofs is increasing. In order to match the printing characteristic using off-press proofing systems gives rise to a different set of problems to the use of proof presses. Most off-press proofing systems are photomechanical, *ie* they rely on a photographic exposure from the halftone films to form a coloured image. There are four main groups of off-press proofing systems:

Laminate membranes on base stock.

Sensitised ink on waterproof stock.

Electrophotographic, based on the electrostatic process.

Direct digital using computer-based data.

The last group is covered later, but the first three categories are the subject of this section. These systems have in common their production by exposure to the halftone films which will be used to make the printing plates. They therefore represent an accurate record of the layout of the page. The position of pictures and text can be judged and measured from these proofs. We have noted that the object of a proof is to match the printing characteristic to be used. We need to examine the methods available when photomechanical proofs are made, to ensure that the proof synthesises this characteristic accurately.

We may remind ourselves that the five main factors in achieving a good match are:

Printing substrate.

Colour of the ink to be used.

Solid printing ink density.

Trapping characteristic.

Tone reproduction.

Printing substrate
Many off-press proofing systems rely on their own special substrate being used. The substrate may be paper or plastic-based. It is usually white but may be pre-coloured to match news or periodical grades of paper. Few off-press proofing systems allow the use of the actual printing stock, unless they demand that the

whole coloured image be transferred from a carrier. All the off-press systems tend to increase the bulk of the receiving stock by adding, to the base substrate, the weight of the laminate image carriers.

Colorant

Conventional proofs, in general, are produced using inks of the same pigment as the press run and hence the problem of metamerism between the two results does not arise. However, in some of the off-press systems the dyes or pigments used are certainly not of the same spectral reflectance characteristics as the ink pigment, even though they may be a close match for colour under standard viewing conditions.

If we could be sure that these proofs were only to be viewed by the client under standard conditions then there would be no problem, but in practice this is no always so. It is unfortunate that such a situation should arise since the use of non-standard viewing conditions within the industry makes colour matching assessment very much more difficult. Nevertheless it is a fact of life we must face and so metamerism must be considered.

Solid ink density

Whilst it is difficult to change the solid ink density in off-press proofing systems, it is comparatively easy to specify in the first place. We are usually using different colorants to the printing inks anyway, so it is possible to use lighter or darker colours to effect density synthesis. Adding black or white pigment to the colorant does not achieve the same end as the colorant's transparency is affected. Many systems provide a choice of colorant density.

Trapping characteristic

Few of the off-press proofing systems allow any control over the trapping characteristic. The lay of the colorants in most proofing systems is dependent on the attributes of the colorant/carrier combination. This is likely to be quite different to the way inks print on the press. Many proofing systems are said to achieve absolute trapping (*ie* have no trapping error) and this is very unlikely to match the trapping characteristic of the press.

Tone reproduction

The first element in matching the tone reproduction of the press is to ensure that the halftone values on the film are transferred accurately to the proof. This is under the control of the exposure used but the exposure may also control the physical performance of the light sensitive element, such as its hardness. If too much exposure is used, the halftone dots will reduce in size in a positive proofing system, but increase in size in a negative-based system.

Exposure may be used to control the effective dot gain in this way. A problem can be experienced in that the press characteristic may evoke differing dot gains for differing dot sizes. Simple exposure variations may not synthesise press dot gains throughout the range of dot sizes. Exposure checks may be made by using a print control target of the type referred to in Chapter 9.

It is fortunate that in many off-press proofing systems the optical dot gain, to be expected when using halftone images, occurs. Optical dot gain will be affected by the smoothness of the substrate and proofs that match high dot gain processes (such as newsprint) may be difficult to achieve. Most off-press proofing systems match generally encountered press dot gains well, however, and this explains the success of these systems in the printing industry.

Other proofing systems

The proofing considered thus far assumes that either halftone colour separation films or plates with which to make proofs are available. Increasingly the colour separations are being scanned into electronic page composition systems, and the separations are available only as data until the final output of films.

Off-press proofing systems have a major disadvantage in the filmless environment – they are made from the halftone separation films. Ink-on-paper proofs also require films, and plates of course. In the filmless environment, the job is represented by a computer data file and it is from this information that the proof must be taken. The reasons for requiring a proof do not change.

We may still require a positional proof, to check the layout of the job, without too much concern for colour accuracy; we may require a contract proof prior to printing, or the provision of a target proof for the press run.

The majority of printing processes require the use of halftone in colour reproductions and the traditional purpose of the proof is to prove the efficacy of those halftone films. Therefore, it is argued, the proof must be a halftone image. This argument overlooks the fact that the halftone is intended to be invisible in the printed reproduction. It can be asserted that the proofing system must simulate the appearance of the final printed result. Exactly how it achieves the match is a matter of academic interest. This is an important consideration for the idea of direct digital colour proofing (DDCP). We have to get from the digital data representing the colour separations, to a synthesis of the printed result as cheaply, and it follows as directly, as possible.

A number of technologies are being exploited for DDCP. These include:

Ink jet.

Thermal wax transfer.

Thermal dye sublimation.

Photography.

Electrophotography.

Some proofing devices are purpose built for the repro industry whilst others are converted from other uses. The repro market is a small one for manufacturers, so their products tend to be expensive. The market for computer peripherals is huge, so products from that area tend to be cheaper. The technical back-up from the two kinds of suppliers needs to be considered, however.

Ink jet
Originally ink jet output was thought to be confined to the area of layout, or positional, proofs to check the relative positioning of text and graphics. Ink jet desktop printers are also supplied for the low resolution output of personal computers.

More complex ink jet proofing machines are available where the colorant is 'sprayed' to the surface of the substrate in minute particles, which may be varied in density by controlling the length of time of the spray. Four ink colours are applied simultaneously through four jets.

Ink jet proofers are attached to relatively simple front-end computers which enable the digital data to be transformed by look-up tables. By this means the appearance of the resultant image can be changed dramatically so as to emulate the appearance of the printed result.

Thermal wax transfer
The majority of thermal wax printers produce coarse, dithered halftones similar to low resolution ink jet printers, with which they compete in the personal computer industry. Typically a roll of wax-coated material in the printer is dispensed with the paper stock. The two pass in contact over a thermal array. This array is selectively heated, under the control of the computer, and wax is transferred from the roll to the paper. Transfer is accomplished one colour at a time, so the paper is passed back and forth four times. The roll is coated with four colours in sequence. The thermal head makes for an upper limit to the resolution that can be achieved, but as a cheap output device for layout proofs the thermal wax printer is adequate.

Thermal sublimation
Originally developed for textile printing, the sublimation process is a relative of thermal wax transfer. It is intrinsically the same mechanical process, but operates at higher temperatures. At these higher temperatures, the colorant is made to vaporise rather than melt. The dye coating is mounted on heat resistant film from which it vaporises to condense almost immediately on a resin-coated receiver sheet. A ceramic thermal head is used and resolutions of around 10 lines per millimetre are possible.

The strength of the transferred colorant is dependent on the temperature of the thermal head and the density can be varied by varying the temperature. A near continuous tone image results and the process has been cited as making second generation originals. The output appears much as a colour photographic print would and the construction of the image only shows under magnification.

Photographic
A number of hardware devices support the exposure of colour photographic stock. The output is continuous tone but there is always difficulty matching the colours of the printing processes. Photographic proofs are not often acceptable as colour matching proofs.

These photographic devices and transparency handling devices are used by suppliers to advertising agencies to output second generation originals which may contain much electronic composite work. The agencies send the transparencies to a number of repro houses for large, multi-publication campaigns.

Electrophotography
Electrophotography, whose name is based on the unfortunate amalgamation of electrostatic and photographic, is based on the electrostatic principles of xerography.

Digital electrophotography is a promising means of making proofs, from data, which can be varied in appearance to match a variety of printing conditions. Electrophotographic proofing devices using digital input are purpose designed machines for proofing and their success will depend to a large extent on their manufacturers' ability to match a number of printing characteristics.

Colour photocopiers are already on the market and their ability to match printing conditions has not been fully tested. In principle it should be possible. The Canon Colour Laser Copier deserves mention here as a device capable of providing positional proofs and designer's visuals. The Canon is a scanning photocopier – it has an input scanner, an image processor, and an output scanner – and between input and output, an interface can provide a means of communicating with a personal computer. This enables designers to input a picture, design a page around it, and output the results for layout approval. The photocopier can be used as a normal photocopier in the meantime. This kind of thinking can be extended to the device providing a viable alternative for short run, short lead time colour printing work.

Chapter 8 *Print control*

The control of the printing process is necessary if any attempt is to be made to produce high quality work consistently. We therefore return to the subject of densitometry. Densitometers are used in the printing industry for the measurement of most stages in the process including film measurement.

Density measurement

From the way we have defined density it is apparent immediately that in order to measure a sample, light must be incident upon it and then the amount of light transmitted or reflected must be measured. If we measure both the transmission through the clear film (or reflection from the paper surface) and transmission through the image (or reflectance from the image) then we can define the transmission factor T, (or reflection factor R). Since

Density = $\log_{10} 1/T$ *or* $\log_{10} 1/R$

it is then possible to calculate density. Fortunately, all densitometers available today are built with electronic circuitry that does this and enables density to be read off directly with no calculation required.

Unfortunately, however, the situation is not quite so simple as it may sound from the preceding paragraph, since the definition of density places no restriction upon the optics of the measuring device. It simply states that density is a function of the ratio between two measurements, not how they should be made. Because of this we normally recognise four distinct types of density measurements, as follows:

> *Specular density.*
>
> *Two types of diffuse density.*
>
> *Doubly diffuse density.*

The need for the different types of measurement arises because of the fact that all materials scatter the light falling upon them to a greater or lesser degree. This means that if a narrow beam of light is shone on to the material the light transmitted or reflected will not generally be contained within a narrow beam but scattered as shown in Figure 8.1.

Fig. 8.1 Light scatter at material surface

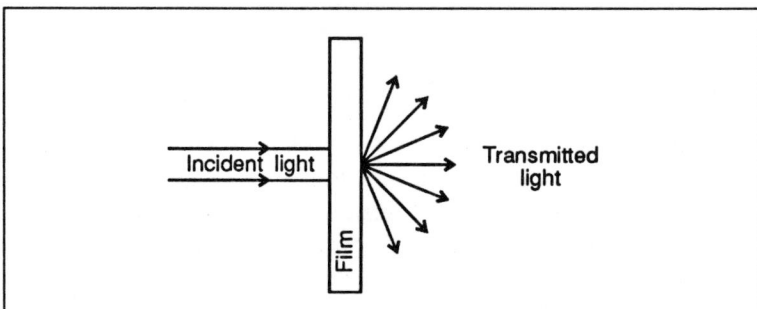

Thus, depending upon where the photocell, which records the amount of light, is placed, the result obtained will be quite different. Figures 8.2 to 8.5 show the various combinations which are generally considered.

In Figure 8.2, the photocell is placed some distance from the film and only the light within a narrow cone is collected by the photocell. This means that the scattered light is largely omitted and hence the density is higher than that obtained with a diffuse density measurement.

Fig. 8.2 Specular density

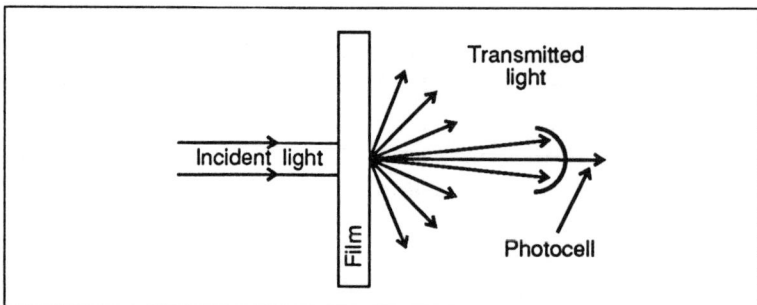

In Figure 8.3, a collecting sphere is placed next to the film to collect all the scattered light as well as that passing through normally.

Fig. 8.3 Diffuse density (1)

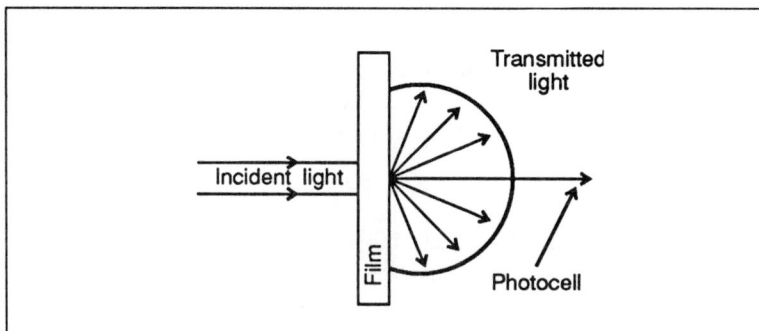

In Figure 8.4, the beam of incident light is diffuse as well as the transmitted light, but the photocell is placed at some distance such that only a narrow cone of the transmitted light is collected. In principle the readings in Figures 8.3 and 8.4 should be identical, although small differences arise in practice.

Fig. 8.4 Diffuse density (2)

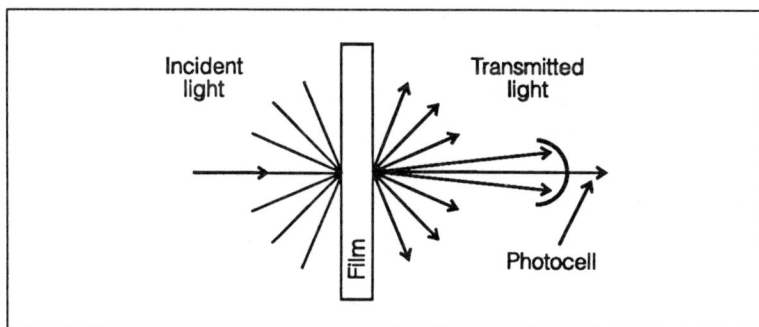

In Figure 8.5, the incident beam is diffused and all the transmitted light is collected.

Fig. 8.5 Double diffuse density

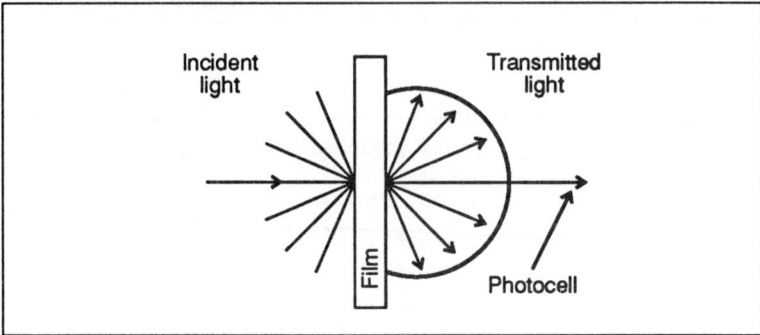

The need for these four different definitions arises because of the way that the film material may be received, or perhaps used in further photographic processes. Consider, for example, two different ways in which a transparency may be viewed. Firstly it may be held up to a window on a dull day or placed on a transparency illuminator. The person viewing it then has his or her eye at some distance from the transparency. This is an example where diffuse density of the second type is likely to be most appropriate since the incident light is diffused and yet the viewer is only receiving a narrow cone of light, that subtended by the eye. If, on the other hand, the transparency is projected via a screen, specular density is likely to be more appropriate since the light in the projector is collimated by lenses and when viewed, a great deal of the scattered light is not seen.

For photographic processes both diffuse density and specular density are important, the former where prints or positives are being made from a negative by contact means, and the latter where enlarged prints are made from the negative. The negative scatters the transmitted light to a limited extent whereas that falling on the negative will be approximately parallel. When more than one piece of film is combined during exposure, as arises in direct screening, this becomes more significant. For a transparency, however, being colour separated in an enlarger, specular density is more appropriate. The same consideration applies to scanners. If a rotating drum scanner is used, the specular density is appropriate. Flat field scanners are generally provided with diffuse lighting systems and diffuse density is more suitable. In practice, however, rotating drum scanners are provided with a means for measuring density *in situ* and flat field scanners are unable to provide density measurements. Since flat field scanners have to be used with a display monitor, the associated software generally provides a means of measurement which is displayed alongside the picture.

Of course, few people actually keep two densitometers, one for measuring diffuse density and one for specular. In general the same instrument will be used for all measurements and commercial instruments do vary somewhat in their optical geometry. It is largely these variations in optical geometry that cause different devices to give different readings, although with transmission instruments the disagreement is generally small.

With reflection measurements we can, in principle, have instruments which measure all four types of density defined earlier. In practice, however, all commercial devices measure a form of diffuse density as shown in Figure 8.6.

Fig. 8.6 Reflection densitometer geometries

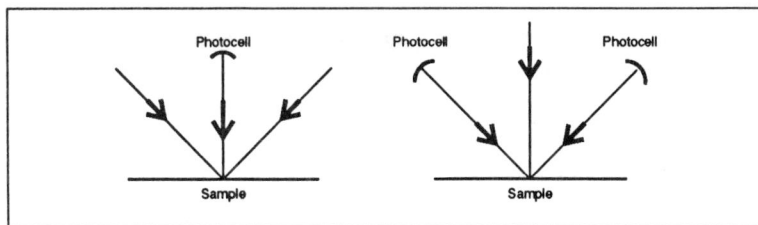

Figure 8.6 shows two common types of reflection densitometer geometries, one which illuminates the sample with light at an angle of 45° to the sample and has a photocell placed perpendicular to that, whilst the other is the reverse of that. This means that the specular component of the reflected light (*ie* the light reflected from the top surface) is largely avoided, although this will depend upon the degree of gloss of the sample; the lower the gloss the greater the amount included. The reason for this is that when viewing a reflection sample the specular light is deliberately avoided and hence we do not require to measure it if we are to obtain meaningful measurements.

155

Many modern reflection densitometers are provided with an intermittent light source which operates only when the densitometer is closed on to the sample to be measured. Battery power is saved in this way. They are fitted in most cases with four filtered photo-diodes as photo-sensors, so that all measurements are taken in one action. The display can either be switched to display each colour channel or all channels at once depending on the design. The instrument is usually fitted with a means of calculation so as to provide all the required measured parameters in one instrument. The model specification (and price) determines the extent of the measuring facilities.

Colour printing

All pressmen appreciate the importance of controlling the printing variables throughout the run length. When the printing press becomes part of a systematic approach to colour reproduction its control from one run to the next is also of significance. The pressman should, therefore, be aware of the influence that press settings, materials, *etc*, can have on the printed result and endeavour to control them.

The variable factors which it is necessary to control to maintain consistent colour reproduction are as follows:

The printed ink density (solid).

Register between colours.

Trapping of one colour on another.

The tone value of printed halftone dots.

All of these variable factors can be influenced by press settings and the materials used.

Controlling solid ink density
Maintaining a consistent solid density (with suitable tolerances) does not prove too difficult a task for the experienced press operator, particularly if they are assisted by a densitometer. In colour reproduction by the halftone process, however, it is vital

that tone values be controlled combined with the solids. This is more difficult to achieve and will be discussed later.

Press settings and materials which can affect the printed solid density include:

Ink feed.

Damp feed.

Level of blanket contamination.

Paper properties, ie *absorbency.*

Controlling register between colours

It is not often appreciated that in the halftone process the colour which is visually perceived from a combination of the process colours can be influenced by the register of these colours to one another. In practice, however, these differences are only slight and do not generally have a significant influence on the final reproduction. What is of greater relevance is the effect which register has on the definition and detail of the reproduction. The press operator will endeavour to maintain as accurate register between colours as is possible. The tolerance will vary depending on the class of work but will normally be in the region of ±0.05mm between colours.

All of the press settings which control the passage of the sheet/ web through the press will effect register, as will the condition and properties of the stock on which printing is taking place. Ambient conditions can also contribute to the variations in register, particularly by affecting the paper properties.

Trapping

The trapping or transfer of one ink on to another, whether it be wet-on-dry or wet-on-wet, is seldom 100%. That is to say, for a given ink film thickness, ink transfer to unprinted paper is greater than to previously printed ink film. This in itself does not present too great a problem. Difficulty is encountered, however, if the trapping characteristics do not remain consistent or when trying to match on a multi-colour production press (wet-on-wet) prints produced on a single colour proof press (wet-on-dry).

Some of the factors affecting trapping are:

Ink film thickness.

Area coverage.

Ink rheology.

Properties of the paper, ie *absorbency.*

Time interval between printings.

Wet-on-wet or wet-on-dry.

It can be seen that when one particular press is being used, control of most of these variable factors should be possible, the main exception being ink rheology which may change during the course of a day as the press inking system warms up and with the take-up of fount.

Halftone dot gain

The term dot gain is an unfortunate one because it implies an error in printing. Whilst it is possible to induce an actual increase in the size of printed halftone dots (by increasing the printing pressure, for example), there is an unavoidable increase in density which takes place at the paper surface. This is often referred to as 'optical dot gain' and its extent is dependent on the roughness of the paper surface. It is caused by surface reflections and ray diffusion at the paper surface. Yule and Nielson described the effect well in 1969, and suggested a factorial correction to the Murray-Davis expression for dot gain. They found that printing on enamelled steel produces very little optical dot gain owing to the hard, impervious surface. Their factor is rarely used nowadays.

Call it what we will, dot gain is a most important factor in halftone printing and can be difficult to control both during the run and from one run to the next. It is made more difficult in that it has to be controlled while maintaining a consistent solid density.

It is unlikely that a production press will be capable of printing dots on to paper without a change in dot size from that which appears on the plate. The dots reproduced will, in most cases, be larger (dot growth) than those on the plate. Sharper (smaller) dots are only likely to occur if the plate wears or is over-exposed, if excessive damp is being run, or if ink piling on the blanket occurs.

The important thing is that the particular characteristics of press, paper and ink, *etc*, remain consistent and are known.

If dot growth is excessive it may be due to a number of factors, including:

> *Excessive ink weight.*
>
> *Ink of low viscosity.*
>
> *Unsuitable blanket.*
>
> *Contaminated blanket.*
>
> *Excessive pressure between plate and blanket.*
>
> *Excessive pressure between paper and blanket.*
>
> *Excessive ink temperature.*

Dot growth will also occur as a result of the dots becoming distorted. This may take the form of slurring (lateral or circumferential) or doubling.

In circumferential slurring, the dots are distorted in the direction of printing. This occurs as a result of the plate moving relative to the blanket or the paper moving relative to the blanket. Over-inking may also produce a slur in the direction of printing.

Slurring and doubling are also caused by:

> *Untrue rolling of cylinders.*
>
> *Wear in gears.*
>
> *Loose blanket or plate.*
>
> *Excessive pressure between cylinders.*
>
> *Movement of sheet out of grippers.*
>
> *Variable tension on web.*

Lateral slurring is less common and occurs as a result of movement of plate or stock in relation to the blanket across the axis of the cylinder. Worn thrust bearings on the plate or blanket cylinders are the most common causes.

In doubling, a second image prints adjacent to the main image, normally at a lower density. It occurs when the ink from a previous printing unit is deposited on the blanket of a following unit via the printed sheet (back trapping). If a deviation in register occurs between the two printing units, the ink is deposited on the blanket in a different position to the previous sheet, resulting in the next sheet printing with a double image.

From this explanation it can be seen that any factor which causes register deviations between printing units can lead to doubling.

Instrumentation and colour control on press run

Once a pass sheet has been obtained at the start of a run this defines the amount of ink required in both solid and halftone areas to match that sheet on all subsequent impressions. Whilst control can be undertaken visually it is advantageous to make use of instruments. Firstly, they offer an objective measurement which can be recorded and used in subsequent jobs, secondly they give an indication of the direction of error when variation occurs and, finally, statistical analysis can be undertaken to properly establish trends and ensure that correction is not being made to the press on the basis of a random error.

For a single ink solid colour, since all that can vary is the ink film thickness, a single measurement of this parameter would suffice. Unfortunately, ink film thickness is not easy to determine absolutely but, in fact, this is not necessary. A change in the ink film thickness causes increased or decreased absorption of the incident light and a measure of this change in light intensity will serve as a control parameter for film thickness and, thus, colour. In fact, it can be shown that for non-turbid media:

$$\text{Density} = \log_{10}(1/\text{Reflectance})$$

is the most appropriate scale since it is more sensitive to changes of thickness than reflectance alone. This is shown graphically in Figure 8.7 where it may be seen that, as the thickness increases, at relatively high densities (above 0.36 approximately) the density scale is more sensitive to changes of ink thickness. Furthermore, for non-opaque colours a greater absolute change in reflectance with ink film thickness occurs where the value is low and thus by choosing to measure changes in reflectance where the value is low (and hence the value of density is high) a reasonably sensitive method of measurement is established. Hence the use of densitometers for colour control in the printing industry. In order to obtain maximum sensitivity from the device it is desirable to measure with light of a narrow spectral bandwidth matched to the absorption peak of the ink being measured, although in practice this is difficult to achieve. Even where the absorption peaks of the inks do not vary from job to job, such as in process colour printing, and specific filters could be nominated, there is the difficulty of obtaining sufficient light with a narrow band instrument.

Fig. 8.7 Density and reflectance as functions of concentration and/or thickness of the colorant layer

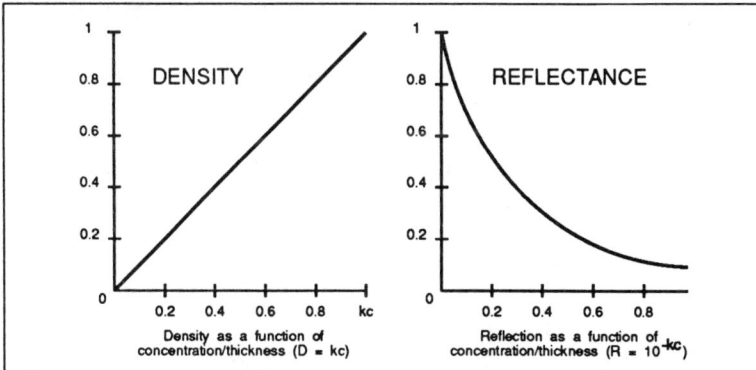

Density as a function of concentration/thickness ($D = kc$)

Reflection as a function of concentration/thickness ($R = 10^{-kc}$)

For this reason densitometers have been traditionally produced with wide band red, green and blue filters for use with process inks. In fact, for inks such as these, any improvement in sensitivity with narrow band devices is likely to prove fairly minimal. The advantages of inter-instrument agreement, by using ANSI Status T filters, are anyway more important these days.

161

From Figure 8.7 it may be seen that at high values of reflectance this scale is more sensitive, but as the values decrease density becomes the more sensitive scale. The exact value at which this change over occurs is:

$$\text{Density} = \log_{10}[\ln(10)] = 0.3622$$

This is shown below.

The gradients are equal when:

$$\frac{-dR}{d(kc)} - \frac{dD}{d(kc)}$$

ie $2.3026\exp(-2.3026\,kc) = 1$

$$10^{-kc} = \frac{1}{2.3026}$$

$$kc = -\log_{10}\frac{1}{2.3026}$$

$$= 0.3622$$

Whilst there can be little doubt that conventional densitometers are a very useful tool in maintaining colour control, they nevertheless have certain disadvantages compared to an ideal device. Quite apart from the problems of obtaining a narrow waveband there is the time taken in removing a sheet, performing measurements and then making corrections. This interval can, on a large sheet, be of the order of 10 minutes or more, during which time 1,000 sheets may have been printed.

Many machine operators (minders) believe they can maintain better control visually, particularly with the help of an accurate proof. Perhaps a more serious technical criticism of densitometers rests on the fact that they are not colour measuring instruments. Whilst the device tells us that an ink film is varying, it is not possible to quote a simple tolerance which is equally applicable to all colours. Thus, the first dilemma of a company installing densitometers is setting the limits for action. This is generally achieved by stating a specific limit (*eg* ±0.05 units) for all colours, but there is no guarantee that these tolerances are optimum. Ideally it is necessary to determine the levels for each colour experimentally, or by using the technique of finding the optimum inking level outlined in Chapter 9. With a true colour measuring device, tolerances are more readily defined. The problem of defining tolerances is particularly serious when colours other than process inks are being printed. If the colour is a very light one it may prove difficult to find a suitable filter for measurement, in order to obtain densities above 0.36, and again the tolerances prove to be a problem to establish.

Thus, we may conclude that densitometers have a valuable role in printing, for establishing variations from a nominal ink film thickness and/or dot area, but that there are distinct limitations. The problem of tolerances is one that requires some degree of pre-calibration. For non-process colours this may prove to be a particular difficulty. It is by no means certain that colorimeters would prove economically viable alternatives as a general rule, although they may be of assistance in certain instances. Modern colorimeters are much more portable than hitherto, although still rather expensive, and much research effort is currently underway to apply them to the printing process. We are just beginning to see them fitted to printing machines as accessories.

On-press measurement systems
The problem of delay between removing sheets from the delivery and making measurements is one that has worried printers for some time, particularly with web presses, and this is an area where some improvements are being made. Pira produced an on-press ink monitor in the late 1960s for letterpress, which was described by Bardouleau in 1970. He showed that the run variations using instrumental control were considerably reduced over those arising solely from visual inspection. Thus, waste can be minimised. With the higher speeds of litho presses it would seem logical that such devices should prove even more valuable. However, there are two factors affecting density with a litho press, both ink and water, and this had tended to retard the development of suitable instrumentation. Although an ink/water monitor was developed by Pira, it was never produced commercially and probably the first available were a family of tack, ink

163

film thickness and damp monitors produced by FOGRA and marketed by Grapho-Metronic. These were open-loop systems but probably, because of cost, have not been readily accepted by the industry. All three units measured physical situations on the press and thus needed some considerable pre-calibration to relate the values to the effect on a running sheet. Effectively each job needed to be calibrated if either ink or paper were changed.

Since then a number of devices have been described which measure the printed density on-line, on a press sheet or web. A Macbeth device was developed in which the sensing heads traverse the web to cover its full width and measurements are initiated using the stroboscopic principle related to the press speed. Because a high power light source (xenon flash) is utilised narrow band filters may be chosen and, in principle, could be of any desired waveband. The data was displayed on a VDU and when pre-set limits were surpassed some form of alarm was initiated. The devices were open-loop and the operator still had the responsibility of deciding what corrective action to take. The development to completely closed-loop operation seems faulted due to our far from complete understanding of ink/water interactions, factors affecting density such as dust and lint on the blanket, and also the long time-constant (hysteresis) of most machines to reach a new equilibrium after adjustments have been made.

These devices, despite their advantages, still do not readily give meaningful tolerances. These would again need to be established in advance of production. Having done this, however, the advantages of real time information seem to be considerable. For certain packaging applications, it would be useful to have colorimetric filters fitted and thereby obtain meaningful tolerance data immediately.

The electronic logic behind these devices is still not totally clear although it seems likely that they are primarily designed for measuring solid areas. The Macbeth device with its traversing heads should be capable, in principle, of measuring halftones also and thereby monitoring dot gain and even secondary colours to assess trapping.

Needless to say, such on-press monitors are likely to prove expensive, and, at present, to be difficult to fit to sheet-fed presses. An alternative approach is that adopted by Heidelberg and MAN Roland with their CPC2 and CC1 units respectively. With these systems a sheet is placed on a measuring table and a variable position densitometer used to measure the colour bars. With both systems the ink duct is initially operated manually from a remote console included in the measuring table but, after this, subsequent sheets are measured relative to pre-set values and the duct settings automatically adjusted to obtain these values. Thus, this is effectively a closed-loop system, although manual override must be incorporated for the reasons discussed earlier.

For speeding up the make-ready procedure there are a number of systems available which scan the area of the negative or positive used to produce the plates, or the plates themselves, and use the data to pre-set the ink duct. This procedure only gives relative settings across the duct length. The absolute level must be adjusted by visual assessment or densitometrically. This, it is claimed, dramatically reduces setting up time of the press.

One perceptual problem with colour control on the press is that of dry-back. A freshly printed ink film has a high proportion of its vehicle and pigment sitting up on the paper surface and producing a relatively glossy finish despite the surface structure of the paper (see Figure 8.8). Once the vehicle has fully penetrated the paper, however, the surface of the ink tends to follow that of the paper and thereby reduces the gloss. Thus, the colour changes as the ink film dries. This means that density readings on a freshly printed sheet change as the ink film dries and, thus, should always be made at the same time interval after printing, or a calibration curve relating density to time produced. Some densitometers manage to reduce this problem quite significantly by cross polarising the incident and reflected beams. Since the specular light reflected from a surface retains its polarisation, it may be eliminated from the measurement. Of course, such devices cannot be used where absolute density values are required but for control purposes they are quite adequate.

165

Fig. 8.8 Typical surface characteristics of wet and dry ink films

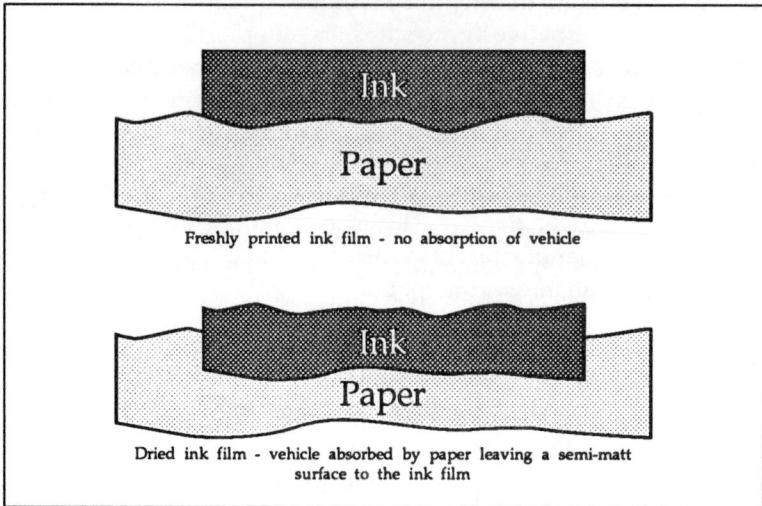

Freshly printed ink film - no absorption of vehicle

Dried ink film - vehicle absorbed by paper leaving a semi-matt surface to the ink film

Finally, it must be stressed again that densitometer control is designed to maintain a specific density. It is only in that context that it is particularly useful. Thus the first task in setting the press is to obtain an acceptable pass sheet. The densitometer, either hand-held or on-press, is then used to measure deviations from this. With the possible exception of process colour work, a colorimeter is essential to specify objectively whether the pass sheet matches the original sample in terms of colour.

However, simply measuring solid ink density is not sufficient. In order to control properly the press sheet it is necessary to measure tone areas (to establish dot gain) and overprints (to evaluate trapping).

Tone areas

Tone areas can be measured for density, and on the basis of this it is possible to establish whether or not dot gain is varying. However, it is difficult if solid densities are varying. In order to determine whether or not dot area is the variable, an equation is commonly used, based on the simple relationship:

$$A = a/A_s$$

where A = light absorption, a = dot area and A_s = solid density

166

When converted to density this gives the equation:

$$D = \log (1 - a(1 - 10^{-Ds})) \qquad \text{(Murray-Davis)}$$

where D = density of the tint and Ds = density of the solid

From this equation we can determine the halftone dot area required for a given density. This is the relationship used by densitometer manufacturers in their instruments when 'dot area' is selected as the measuring mode.

Unfortunately, due to light scattering in the paper, the resultant area is not an accurate one, though quite acceptable for control purposes. If corrected to take account of scatter we obtain an equation of the form:

$$D = n \log (1 - a (1 - 10^{-Ds/n})) \qquad \text{(Yule-Nielson)}$$

where n varies between 1 and 3 depending upon the material being printed, and is the correction factor referred to earlier.

For practical control purposes, the Murray-Davis equation is probably adequate and is the one commonly used.

Trap
Measurement of trapping generally relies on additivity being observed, and as we are aware this is rarely true. Once again, however, as a control tool it is quite satisfactory and is established by measuring the density of the two ink films, then the over-printed ink, all through the same densitometer filter.

The ratio $(D_{ov} - D_1)/D_2$ should be constant throughout the run, where D_1 is the density of the first printed colour, D_2 is the density of the second printed colour, and D_{ov} is the density of the over-printed pair. The densitometer filter is that which is suitable for the second printed colour in all cases.

Thus, we can see from this discussion that a densitometer may be used for measuring dot gain, solid density and trapping. This is very useful control information and can certainly be used on a comparison basis between printed sheets and between printed

167

jobs. In order to make optimum use of the densitometer it is
desirable to undertake a proper statistical analysis of the data.

Measurement targets (control bars)

As has been stated above, a densitometer can be usefully employed in multicolour work for controlling or monitoring:

Tone reproduction.

Dot size.

Solid density.

Changes in grey balance.

Trapping.

While these aspects of a printed reproduction require control, it is
also only of slightly less value to also be able to assess other print
characteristics such as:

Register between superimposed print images.

Resolution.

Print sharpness.

Slur.

Dot doubling.

Although these are not all measurable with a densitometer, they
can be given a relative value by means of special patterns or
targets incorporated close to the reproduced picture as part of the
printed image.

Advantages of process control strips

With the exception of the colour parameters of a print, the use of
a suitable densitometer and a specially designed target is very
desirable because they are able to indicate changes in the printing
performance of the press or materials far more easily than is
possible visually. The ability to detect a change quickly can also
lead to an early remedy and savings in time, effort and money.

By keeping the process within acceptable limits of quality it is possible to obtain:

Optimum ink performance - one that gives a maximum tonal range at an optimum density.

A reduction in ink consumption.

Avoidance of set-off and/or marking

Better control of colour balance with the use of a suitable target.

Better assessment of the characteristics of the press behaviour through the detection of slur, ink distribution, and register, or the determination of the best possible impression setting.

Screen, plate, paper, blanket, and ink performance assessment and specification more accurately than by viewing areas in a halftone reproduction

Better correlation between proof and production print.

Ultimately, by being able to better quantify the quality of the print, we can statistically process such data to remove the random variables of the press and materials, and thereby produce standards for printing.

Critical print characteristics and special targets
Tone reproduction

Tone reproduction control in conventional halftone printing usually relies on providing several halftone dot areas, each area being separate and of a regular dot value. This forms a mini-halftone scale.

By providing at least three appropriate sizes of dot, a three-point tone scale will indicate if all tones in the halftone printed image are being reproduced in tolerance. Usually the three values chosen are tones of 25% or less, approximately 50% and/or 75% tone, and a solid patch. It is common to duplicate the use of the middle and/or the high tone value for monitoring changes in dot size as noted in the next section.

If all the tones are printing correctly, *ie* without missing highlight dots, filling in shadow tones, or either showing excessive dot size changes, then the tone reproduction of the print is likely to be fairly satisfactory. The assessment of the reproduction of the target is best carried out with the densitometer, although these tones are also readily evaluated by the experienced eye.

Optimum tone reproduction determination requires a more lengthy process of increasing the ink weight in repeatable stages and measuring the density difference between the tone (65%, 75% or 85% tint) and the solid target. By a single calculation, *ie*

$$\frac{\text{Density}_{\text{solid}} - \text{Density}_{\text{halftone}}}{\text{Density}_{\text{solid}}}$$

it is possible to derive a value called 'print contrast'. A succession of these calculations at different ink weights can then be plotted on a graph.

Fig. 8.9 Print contrast

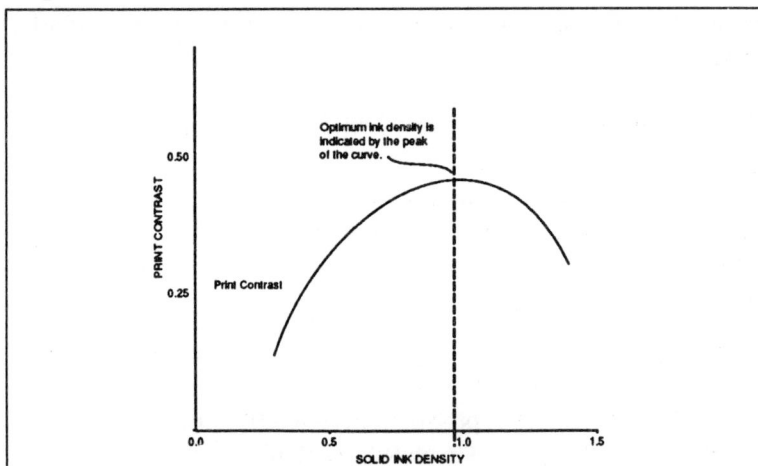

The extended nature of this test means that this is not really a production quality control method. However, it is one of several very valid procedures for establishing optimum tone reproduction, and is particularly useful during the installation of a new press. (See also Chapter 9 for a full implementation procedure for finding optimum inking using print contrast.)

Dot size measurement

Tone control is really the culmination of a series of dot density measurements. Currently, the methods of monitoring tone density changes rely on one of four methods:

Screen area density in relation to solid area density.

Normal screen area density in relation to coarse or fine screen area densities.

Lines of tapering width (radial star), or varying distance.

The methods vary between manufacturers of the control bars and each supplies instructions for their use.

Primarily, the aim here is to provide a regular pattern of halftone dots or lines that are especially likely to change their optical density with different ink weights, viscosities, tack, or press pressures. The most suitable halftone value for this work is thought to be in the 75% tone region, although not all QC strip manufacturers subscribe to this. The coarse screen or solid patch density is a reference point. Generally by noting the difference between the solid (or coarse) density and the tone density, a density value is obtained from which any deviation indicates a likely and proportional change in the tone value.

Various factors, both in terms of unavoidable press influences and also more complicated optical factors, make basically simple targets and their density values rather difficult to interpret at times, but these phenomena will be described later.

If the targets are regularly used, it will soon be evident what correction is required in the film positives or negatives in order to produce a satisfactory reproduction, or what change of dot size in the target is accompanied by a density change in the print.

Solid density

Even in single colour work, a solid printing area can be used for checking the maximum printed ink film thickness for optimum reproduction and setting the ink duct for even inking, in addition to acting as a possible reference element for any dot density monitoring system incorporated in the QC strip.

In colour work, solid areas for each colour printed are further needed to assess the print characteristics of overlapping ink films - the secondary colours - whether as an indicator of the transparency of the overprinting ink (it has to be for colour printing to succeed), or as a means of evaluating the thickness of ink film deposited by superimposed colours.

Grey balance

The theory is that it should be possible to print the 'colour' black with solid overprints of cyan, yellow and magenta, since printing is a subtractive process. Following on from this, 50% tone areas of these three-colour inks, when overlying each other on white paper, should appear as a neutral grey. However, owing to certain deficiencies in the ink, discussed in Chapter 4, the result is normally a brown tint. In practice, the grey balance is achieved with a combination of approximately 50% yellow, 50% magenta and 70% cyan overprints.

Provided the separation film positives have been balanced using the proportional tone correction factor required to achieve a grey balance, and that the printed density has been taken account of, good tone reproduction should follow. But, due to certain variabilities in platemaking and printing-down, it is possible to lose the balance. To guard against this loss, it is advisable to print targets made to the relative proportions so that a neutral grey tint occurs when they are printed over one another. Several grey steps of, say, 25%, 50% and 75% could be composed provided their component colours were in the same relative proportions. This mini grey scale would then be subject to both the platemaking and printing variables, and would indicate if the grey balance of the printed image was being maintained. Further, the press should be run to keep the inks in balance.

It is important to retain this balance if the tone reproduction of the print is to remain true. Correct grey balance is visually more important than precise colour correction. Colour error can vary in quite large degrees, but a grey balance deficiency is visually more apparent and therefore serious.

It should be mentioned that the targets are judged by eye by most printers. Densitometers are rarely used for grey balance control. There are two main reasons for this.

Firstly, the target is only of use in wet-on-wet printing since ink weight changes must be made immediately if the colour balance needs correcting. Fortunately, the human eye is quite good at discerning whether a grey is neutral or has a colour bias. A visual assessment of the target can, therefore, be made as quickly as any instrumental system could.

Secondly, one of the great assets of using a densitometer is that the readings shown are not subjective, and that the results can be recorded. Unfortunately, the densitometer cannot provide true colour values, and colour density values through red, green and blue filters take time. Further, the available methods for plotting the greys from colour density values are also very time consuming, with the resulting graphical plots often relating poorly to the visual effect.

Other targets
Other types of targets in the printing control range are also assessed visually. Basically they can be categorised as:

Plate exposure control devices.

Resolution guides.

Slur guides.

QC (quality control) strips
The targets for the control of print characteristics are usually grouped together in the form of a narrow strip, for ease of handling and to economise on space when printed on the sheet (see Plate 7).

One of the critical considerations when using QC strips is their position on the printed sheet. This has to be in a place where it will experience printing conditions representative of the whole of the plate, yet can be trimmed off easily during the finishing operations. Ideally, this is within the first third of the sheet, running across the press (according to the German Graphic Arts

Research Institute (FOGRA)). In magazine printing it is often placed in the gutters between pages where the paper is folded, and finally trimmed.

Other properties of the strip which will be described more fully are:

Measuring area size.

Importance of the measuring area surround (and viewing conditions for grey balance assessment).

Repetition of the critical targets.

Functions of groups of targets.

Dimensional stability.

Susceptibility to fogging and ageing.

Resolution.

Effect of diffusion sheet.

The practical aspects of printing them down and using them.

Limitations

Gloss

Since the targets will often be measured straight off the press the ink films will be wet. The high gloss which is associated with one wet film gives a high degree of first surface reflection. The densitometer, with its illuminating source, would indicate these gloss prints to have a higher density than that of the dry ink film. This phenomenon can be counteracted by illuminating the sample with light that has little or no light radiation at any direction other than the direction from which it comes.

Ordinary light that comes from the sun or from incandescent (tungsten filament) bulbs consists of light waves with a complex mixture of vibrations lying in all possible directions crosswise to the line of travel (unpolarised light). If the vibrations are in one direction, the light is linearly polarised.

By using a polarised light source the amount of light reflected by the surface (first surface reflection) can be virtually eliminated. The readings of a densitometer fitted with such a set of filters are similar whether the ink is wet or dry.

Proportionality failure

The purpose of the pigment in an ink is to absorb light of various colours selectively, while the paper's function is to reflect the remainder. However, light passing through the ink layer is reflected in all directions by the paper, and some of it is reflected back toward the paper by the upper surface of the ink film to be reflected again by the paper (see Figure 8.10). Eventually it either emerges or is absorbed.

Fig. 8.10 Proportionality failure

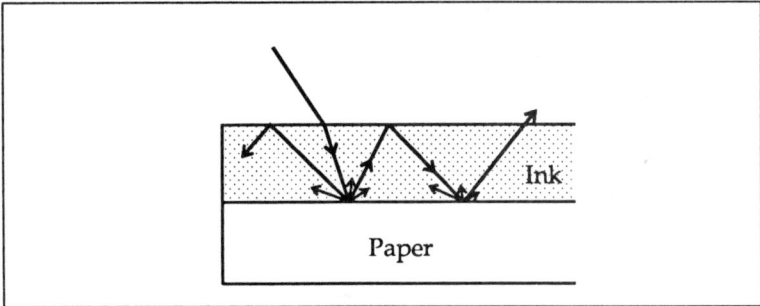

Because of this, the apparent density of the ink layer is greater, and also not directly proportional to the amount of ink present. This is called proportionality failure.

The degree of proportionality failure is particularly noticeable in halftone work when coarse screen tones are compared with fine screen tones, it being worse for coarse tones. With fine screen tones, the light penetrates far enough into the paper for the dot pattern to become diffused before being reflected and passed out through the ink layer or between the halftone dots. Proportionality failure is particularly important in considering one application of densitometry, that of halftone dot size control.

Without the complications of these multiple internal reflections of light, it would be easier to relate halftone density to tone value. In order to obtain a better correlation of dot size to density we have to use more refined methods such as:

Mathematically adapting the density reading using the formula derived by Yule and Nielsen.

Using video scanning, image analysing techniques as employed in the Pira or Beauvac Electronic Planimeter.

Additivity failure

Multicolour printing involves the overlapping of several ink films, and it is important that an overprinting ink is accepted equally well on already printed areas of the sheet as it is on unprinted areas. The ability of the overprinting ink to do this is known as its trapping properties. Poor trapping means that less overprint ink is laid down on the printed area so that there is a colour bias towards that of the underlying ink.

To measure trapping, reflection density readings are usually taken from the solid areas of a QC strip through the colour filter that makes the top colour look black.

Fig. 8.11 Trapping viewed sectionally through ink films

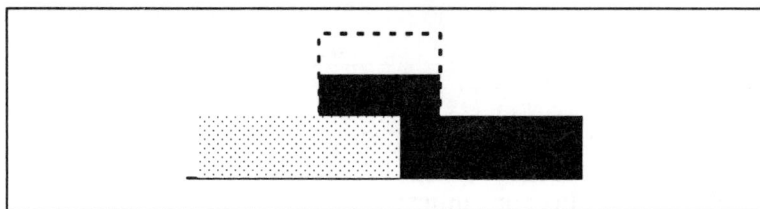

By calculating

$$\frac{D_{ov} - D_1}{D_2} \ (\times\ 100)$$

where: D_1 = density of bottom ink colour
D_2 = density of top ink colour
D_{ov} = density of overprint

we often find that the result is less than 100%. That is to say, the density of the two overlapping ink films is rarely $D_1 + D_2$. This

176

discrepancy is commonly taken to be the result of poor ink trapping, although to be more accurate the phenomenon should be called the additivity failure property of the ink, since trapping is a contributor to this property rather than the sole cause of it.

The following are the main causes of additivity failure:

First surface reflection.

Multiple internal reflection.

Ink opacity (lack of transparency).

Poor trapping.

Back-transfer (partial removal of the underlying ink film by the adhesive and tacky nature of the second ink film, and/or contamination by mixing of the pigments of wet inks at their interface).

Spectral characteristics.

Halftone structure.

*Light scatter in paper.**

(* Light scatter in the paper is not often a serious contributor to additivity failure since the effect is largely nullified by the common practice of zeroing the densitometer on the paper before ink densities are measured. However, combined with the halftone structure it does become a problem.)

Additional applications of test targets

Less usual applications of test strips or forms that often require the use of densitometers can be listed as:

Determination of optimum ink film thickness (measured as ink density).

Examination of the effect of hard or soft packing of the blanket.

Determination of the best possible impression setting.

Checking of blanket quality.

Control of measures to make proof and machine print identical.

Determination of tonal value reproduction characteristics of inks.

Checking the behaviour of papers in relation to inks.

Determination of ghosting, halftone unevenness, resolution and print sharpness performance of the press.

Determination or tonal value reproduction characteristics of different types of printing plates and printing down methods. Determination of ink density changes on drying (dry-back).

Chapter 9 *Specifications for reproduction and platemaking*

The reasons for specifications

Printers and their customers often disagree whether any given reproduction is 'good' or even acceptable, and one of the prime reasons for this is poor communication. The platemaker may be presented with originals which are unsuitable for reproduction and must be distorted in some way to meet the requirements of the printing process. When the customer is presented with the finished print, it is quite likely that the result is not what was wanted. The customer may see a proof before the printer receives the plates, and accept the distortions in the reproduction revealed by it, but unless the proof is a reasonable facsimile of the production sheet then further distortions may occur which again displease the customer. Whilst proofers must accept some of the blame for this situation, unless the reproduction characteristics of the press are made known to them by the printers they cannot be expected to provide a facsimile of the print. Even if good originals are presented which enable the platemaker and printer to know what is required of them this lack of information can still produce problems. A well defined and accurate specification of printing characteristics and requirements will enable the standard of reproduction to be controlled and make the finished product predictable, the first requirement of any reproduction system. If the output is variable for a given input there is little chance of producing any desired result.

The economic argument for a print specification is also very strong. Apart from the savings due to the reduction of reprints for rejected work, the time saved on machine trying to match proofs not produced to suit the press characteristics will be significant. Publishers need to know that film separations for advertising pages, which may come from different sources, are made to a common specification, and that the specification is that of their

179

press. For the platemaker, having a specification to work to makes the work more precise and with modern scanners, or exposure and developer controls, it is not difficult to meet the specifications. This reduces the necessity for remakes considerably, with a consequent saving of time and materials. There is also improved goodwill between all parties concerned with the obvious benefits for each of them.

Factors affecting the specification

In order to produce a satisfactory specification it is necessary to investigate the output of a printing press and find an objective method of defining the various parameters. It is necessary to find nine variables to ensure that the print characteristic is properly defined. Importantly, the proof should be made to emulate those same print variables in order to be valid. The proof should match the print characteristic in order that the press make-ready time be reduced to a minimum. These nine variables are:

> *Paper.*
>
> *Ink colour.*
>
> *Ink sequence.*
>
> *Ink film thickness.*
>
> *Dot gain.*
>
> *Grey balance.*
>
> *Black printer characteristic.*
>
> *Trapping.*
>
> *Plate transfer characteristic.*

A short discussion of each of these variables follows.

Paper
The paper type, particularly its surface properties, is fundamental to the appearance of the finished printed reproduction. Paper is generally selected for its utility in an application. Newspapers, for

example, only have a short life, so the keeping properties of newsprint are not important. Books have to 'work' as books. They have to be subjected to handling, opening and closing, and may be kept for years, if not decades. The feel of the finished product is important, so an expensive glossy magazine has to feel expensive (and glossy).

For the purposes of the print specification, the paper type may be classified in one of several groups. To be pedantic, the same paper type made in batches may differ from batch to batch, but differences in the appearance of the print may be in tolerance. Generalised groups would be matt coated, blade coated, gloss art, calendered, and so on. In the case of web printing, only a limited number of paper stocks may be routinely used.

Ink colour

The colour of inks is discussed fully in Chapters 2 and 4. It should be borne in mind that the ink has many important properties to satisfy before its colour is considered. It has to stay on the paper, it has to dry on the paper, and it has to be printable. In the vast majority of cases, the ink conforms to a colour specification such as BS 4666, and is purchased ready to print. Whilst BS 4666 is a description of colour, tolerances in the standard mean that it is not a unique description. Several examples of ink sets conforming with BS 4666 may be noticeably different in appearance when compared with each other, but still be in tolerance. Further, the same ink applied to different paper stocks may differ significantly in appearance. The make, type and identification number of inks used should be listed in a printing specification.

Ink sequence

The printing order of the colours varies considerably in the printing industry. It is chosen for a number of reasons, not the least of which is fashion. If inks were ideal, *ie* if they were pure colours, were transparent, had good adhesion to both paper and previously printed ink surfaces, and had ideal surface reflectance properties, then the colour printing order would not be important from the appearance point of view. Inks are not ideal, so the colour sequence is important. There is no preferred order from the academic point of view, though many printers swear that there is only one correct colour sequence – their's. The print specification should state what the colour sequence is, if only so that the proof can be made in the same sequence. A logical colour sequence may be that the colours are printed in order of opacity (lack of transparency), the most opaque being printed first. Note that the most transparent colour ink is not necessarily the lightest colour ink.

Ink film thickness

The amount of ink printed is going to affect its appearance considerably. If a newly opened tin of ink is viewed, its colour varies significantly from the finished printed colour. In the tin, the ink has a thickness as deep as the tin. If the ink has some opacity, then that opacity will increase as the thickness of application is varied. The colour will change with thickness, and the ability of the ink to spread out of the image areas will increase with increased ink film thickness (see *Dot gain* below).

Dot gain

Dot gain is the growth of the halftone dot size under the impression pressure of the printing process. The printing pressure is significant and the ink, at transfer time, is fluid. As a consequence it spreads out from all image boundaries. A degree of dot gain is normal but it will become excessive if there is too much ink, if the pressure is too great, or if the ink is too fluid. Dot gain on rough surface papers is likely to be more than that on smooth papers if only because the printing pressure is generally higher for rougher papers. Even if we could be certain that no mechanical dot gain was taking place on rougher papers, we would find that dot gain, when measured with a densitometer, was apparent. This dot gain, sometimes called optical dot gain, is due to complex reflections at the surface of the paper and should not be confused with mechanical dot gain, though its effect on appearance is the same.

Grey balance

We have noted in Chapter 4 that the accurate reproduction of greys is important in colour reproduction. Since printing equal dot areas of the three primary ink colours does not produce a neutral grey, then a correction must be made to the dot areas printed in greys and colours that have a grey component. This is grey balance correction. It must be determined for each printing condition. Note that if different dot areas produce differing dot gains, then the correction applied is dependent also on dot gain at the dot size concerned. A method must be found of specifying the grey balance in spite of, or including an allowance for, the relevant dot gains – no matter what the cause of those dot gains.

Black printer characteristic

The need for the black printer is necessitated by the application of grey balance, to increase the range of darknesses in the printed reproduction, and to increase the colour gamut. Since the introduction of the black enables darker colours to be reproduced in a number of ways, there is no uniquely correct amount of black that can be specified in the printed result. It is very complex to specify a black printer in terms that are meaningful and checkable. When UCR or GCR are specified the problem is exacerbated. In practice the black printer is often found by heuristic means.

Trapping

Trapping performance is intimately bound up in the properties of the printing inks selected. The effect of trapping is wholly dependent on the colour sequence used. There is anecdotal evidence that the selection of an ink colour sequence depends on the trapping performance of the inks and the importance of secondary colours in the printed products. If the printing order of the yellow and magenta is reversed, then a red will change its hue from orange red to magenta red or *vice versa*. The extent of the change will depend on the difference in trapping performance of yellow on magenta, and magenta on yellow inks.

Plate transfer characteristic

The importance of getting a known dot size on film to transfer to the plate with the same dot size should be fairly obvious. The transfer is dependent on the plate exposure time, exposure lamp geometry, film quality, and plate surface topology. The use of a microline target or a Stoufer wedge will ensure the consistent transfer of known quality film. The step numbers at which the target design is affected can be used as warning of a change in performance, or as an indication of conformance.

Producing the specification

To produce the printing specification, we have to accept various 'givens'. The type of paper will be determined by the designer. The type of ink will be determined by the printer, in most cases, and this choice will have determined the colour of the ink also. The intermediate materials choices – films, plates, and chemicals – will also have been determined in most cases. What we are interested in, in defining a specification, is the behaviour of the materials when they act together in the whole process.

We can identify the objective of colour reproduction as producing an acceptable copy, or copies, of a coloured original. The means by which we will do so is colour printing. Colour printing is a variable process, and nothing of a fixed quality can be produced via a variable process, *ergo* we must fix the variable. It could be argued that the term 'acceptable copy' is itself a variable. But if the acceptability of the reproduction is to be judged by a comparison between the original and the reproduction, we should remember that the eye is very reliable in this mode – given standardised viewing conditions.

The advantages of fixing the printing variable are great. If the print parameters are fixed, we can analyse them for their effect on the appearance of the reproduction, and use the other great variable, the reproduction process, to correct for unwanted effects in the appearance of the resultant reproduction. This is exactly the approach taken in the photographic industry and the following approach has much in common with that method.

Optimum inking level

The most fundamental question is, how much ink should be presented to the paper? It should not be so little that the print is light, and difficult to view. It should not be so much that it will not dry, or that the halftone structure is 'filled in'. Somewhere between these two extremes, then, is an optimum level. We have expressed this as ink film thickness, and have noted elsewhere that ink film thickness, the mass of ink used per unit area, is measurable by use of a densitometer. It seems an obvious starting point to print a series of differing ink film thicknesses and to analyse what differences in appearance, and measurement, occur as a result.

We note that if the ink film thickness is too low, there is a reduction of colourfulness. If the ink film thickness is too great, then there is also a reduction in colourfulness, the ink colour becomes dirty – we would say – dark. Optimum ink level will result, in the first instance, in optimum colourfulness.

The rheological properties of the ink are important in the way halftone values are printed. Too much ink will result in the halftone areas filling in, resulting in apparently solid ink areas where there should be halftone tints, and a reduction in contrast between halftones (especially dense tones) and solids. Less apparent is that too low an inking level will reduce the intensity of a solid area more than an area of halftone leading to a reduction in contrast. Optimum ink level will also result, therefore, in an optimum contrast between solid and halftone areas.

The optimum inking level is not something that is likely to change, and is therefore best found during the installation of a new printing machine. Factors that will effect optimum inking level are press wear, printing temperature, paper surface, and the ink set used. If any of these factors change then, ideally, a new test should be carried out. One advantage of testing a newly installed press is that the test can be repeated at a later date to determine the condition of the press, and whether maintenance is required.

To find the optimum inking level, for a given ink set and paper, it is necessary to find both the optimum colourfulness and the optimum contrast. The procedure to do this is as follows:

> Prepare a plate containing a solid area and a halftone tint area of about 80% and 50% for each colour. A grey scale is ideal for this test, but many control target bars contain the required steps.

185

Run the plates on the press at a number of different inking levels. This may be achieved by running the press at a high inking level, lifting the ink rollers and allowing the press to run on. The ink will slowly run out producing prints with successively less ink film thickness.

Take sheets from the printed stack with reducing amounts of ink printed and measure these with a densitometer recording the values for future analysis. The required measurements are:

the solid density measured through the correct (complementary) colour filter for that colour ink (D_{SOLID});

the 80% tint density measured through the correct (complementary) colour filter for that colour ink (D_{TINT});

the density of the solid measured through the colour filter which gives the lowest density reading for the solid – this will not be the complementary colour filter (D_{LOW}).

Many modern densitometers will allow the readings to be transmitted to a computer and entered in a spreadsheet, which is convenient. Otherwise the readings can be manually entered to a computer or recorded for later manual analysis. Note that it is unnecessary to gather data for the black colour contrast, as the black is a non-colorant ink.

From the recorded densities of each sample evaluate each parameter as follows:

$$\text{colour contrast} = 10^{-D_{LOW}} - 10^{-D_{HIGH}} \times 100$$

$$\text{print contrast} = D_{SOLID} - D_{TINT} / D_{SOLID} \times 100$$

Record the colour contrast and print contrast for each solid density in a table such as Table 9.1.

Table 9.1

Colour contrast and print contrast measurements		
Magenta		
Ink density	**Colour contrast**	**Print contrast**
0.8	43	65
1.0	45.5	67
1.2	47	68
1.4	42	66
1.6	33	63
1.8	13	57

Prepare a graph and plot the values for colour contrast and print contrast against solid density (see Figure 9.1). It is most convenient to produce a graph for each colour of ink. It will be noted that most of the plots result in curves for both functions.

Fig. 9.1 Plot used to find optimum ink density

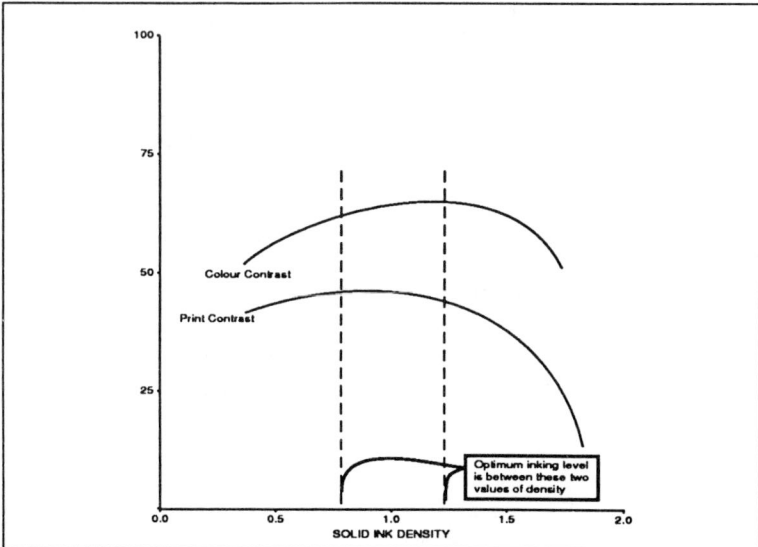

The optimum inking level is indicated on the graphs by the apogee, the highest point, of the curves. These points on the curves represent the maximum colour contrast and the maximum print contrast of the ink, on the paper, printed on the press and at the temperature at which the tests were made. So by finding this point on the curve, the solid density on the x-axis below that point is the density of the optimum inking level.

For each of the samples, measure the 50% halftone area. This is best measured using the halftone scale on the densitometer if it is provided with one. Otherwise the straight density can be recorded and the halftone dot area evaluated according to the Murray-Davis expression:

$$\text{Effective dot area} = 10^{-D50\% \, TINT} - 1 \, / \, 10^{-D_{SOLID}}$$

Plot the dot area on the appropriate graph and note the dot gain that results from the use of the optimum inking level.

Fig. 9.2 Completed optimum inking plots for a magenta ink

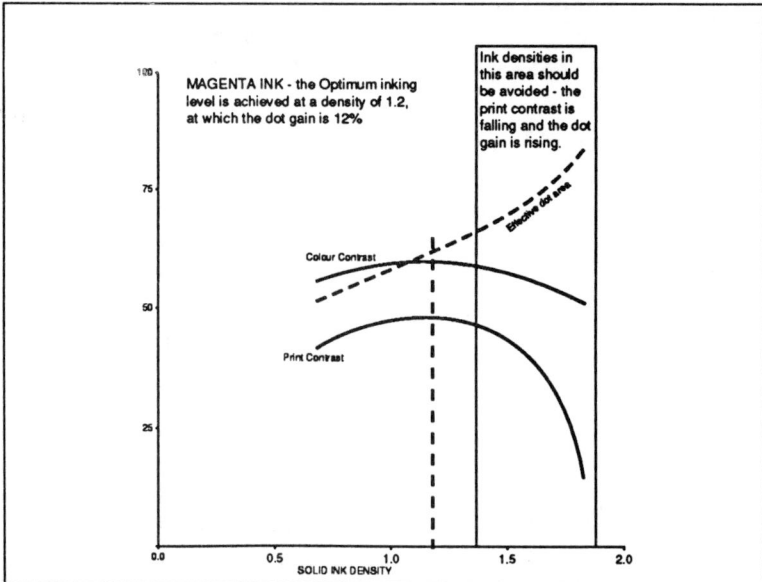

Notes when determining optimum inking level:

> One graph is required for each colour. This makes them much easier to construct and to read off the correct values.

> It may be that the print contrast and the colour contrast call for a different optimum ink density. In this case the density chosen should be a compromise between the two, in such a way that the differences are split. A 'flatter' curve in one of the parameters indicates a greater tolerance, so the choice should be closer to the apogee of the sharper curve.

> There is no point in plotting a colour contrast curve for the black printer – it has no colour! The print contrast curve alone should be used to find the optimum ink density.

> Arbitrary tolerances may be adopted but it makes sense to avoid highly sloped areas of the curves. Asymmetric tolerances may be used (*eg* +0.05; -0.10).

> When the ink density tolerances have been established, then read off the dot gain tolerances.

Determination of the qualitative parameters

Having found the optimum inking level it remains to determine the reproduction factors which have an effect on the appearance of the printed work. Note that the optimum inking levels are now targets for all future printing carried out on the tested press using the tested inks on the tested paper. The use of target colour bars is a means to check the future compliance to the optimum inking level by measuring both the solid ink density and the dot gain. In checking these parameters, the dot gain is the most important of the two.

A printing test form is required in which the required measurement points are conveniently laid out, preferably in a rational order. These are available from sources such as the Rochester Institute of Technology (RIT), scanner manufacturers, and commercial suppliers. They can also be produced on modern reproduction equipment, desktop computers, and even by manually planning the elements from screen tints. Examples are shown in the colour plates.

The factors which are required from the test form are:

Grey balance.

Tone reproduction characteristic.

Maximum and minimum halftone values which are printable.

Maximum printed density of three colours as a grey.

Maximum printed density of four colours as a grey.

Primary and secondary colour gamut.

In the case of the colour gamuts, it would nice to think that these would provide meaningful information for setting up colour correction on a scanner. Whilst they give valuable appearance data, and indicate the sort of values necessary for the laying of mechanical tints, they do not yet provide data which is easy to use rationally.

It is important that the test form is printed as normally as possible. It is usually the case that a printer confronted with the test form will adopt a careful, almost reverent, attitude to printing such a piece. The print test form plates should contain a target bar, and the optimum ink densities, found above, should be adhered to and checked across the whole of the printed area.

The grey balance
We have determined that the greys are important in the reproduction because the eye is able to determine that a colour is, or is not, a grey. It should not then be surprising that the grey samples in the test form are selected by visual means. The greys are selected and the halftone values of yellow, magenta, and cyan are read from the scales, or the test form film is measured to find the dot areas for each colour. The grey samples are measured with a reflection densitometer measuring through the 'visual' or non-colour filter (see also Plates 4 and 5).

190

Fig. 9.3 Halftone values for the three-colour grey test target

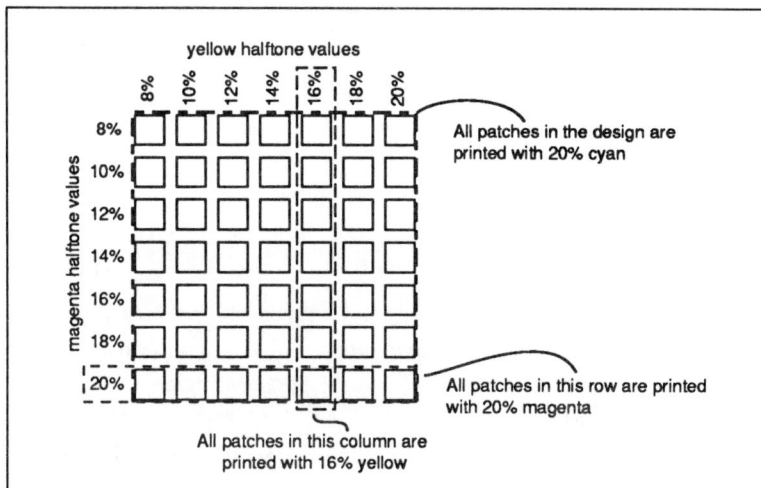

The results are tabulated as shown in Table 9.2.

Table 9.2

Grey balance values

Cyan	Magenta	Yellow	Print density
0	0	0	0.00
5	3	3	0.05
10	7	7	0.09
20	14	14	0.22
30	25	25	0.35
40	35	35	0.48
50	44	44	0.61
60	54	54	0.75
70	64	64	0.90
80	74	74	1.02
90	83	83	1.12
100	88	88	1.20

Notes

In selecting the grey from the samples on the test form some care is needed. It is easy to do – but it is easy to do it wrongly.

Normal, or greater, reading distances should be adopted. If the test form is viewed at too close a distance, the eye will discern the component colours from which the grey patches are constructed. Increasing the viewing distance ensures that only the 'resultant' colour – grey – is seen.

Most supplied test forms have the grey patches printed with a black surround. Be sure that the selection of grey is not made on the basis of equal density to the black surround. It is often helpful to have a copy of the test form printed in three colours only (*ie* no black), so that the greys are surrounded with white.

Do not mark the selected grey patch with a pen or pencil to indicate the grey choice, it has to measured with a densitometer.

The selection of greys should be made under the specified illumination conditions. Failing such a specification, and for preference anyway, use viewing conditions conforming to ISO 3664.

The tone reproduction characteristic
The densities of the grey samples provide a means to determine the tone reproduction characteristic of the printing conditions. By using this information in a meaningful way, we can determine the necessary corrective factor to be applied to the colour separation stage of the process. It is important to realise what has been achieved by this comparatively simple exercise.

The industry has for some time been aware of the importance of dot gain, that dot gain is implicit in the printing process, and that it is implicit in the halftone process. What has not been so obvious is what can be done with a knowledge of the dot gain. It is logical that a high dot gain value should be allowed for in the halftone films, but by reducing the halftone value in one area it is often

192

overlooked that there is dot gain at the new value. What should be done about that?

In order to find the dot gain, it is necessary to measure the density of a halftone area, then manipulate the readings (as above in the Murray-Davis equation). The importance of dot gain is that it has an effect on the appearance of the printed result. The density measurement tells us indirectly just what that effect in appearance is, and more, tells us what it is at the number of points in the scale at which a measurement was taken.

Since it is three-colour greys that have been measured, the densities taken provide data which is meaningful in terms of dot gain, trapping, additivity failure, proportionality failure and pick-off, throughout the range of halftones – a formidable set of data.

In other words, the selection of grey samples and their density measurement tells us all the important things we need to know. In order to use this information effectively we need to express it in such a way that it can be used as a corrective in the generation of the colour separations.

Maximum and minimum halftone values
Inspection of the areas containing the smallest and the largest halftone dots, using a magnifier, will readily reveal the dot sizes which cannot be printed. If the grey balance test area extends into these values, that set of data will also indicate the points at which no change in printed density occurs.

A test piece can be envisaged that will indicate independently both the maximum and minimum dot sizes and the finest halftone screen that can be printed under the tested conditions (see Figure 9.4).

Fig. 9.4 Test targets to find the maximum and minimum printable halftone dot sizes at different screen rulings

SCREEN RULINGS							SCREEN RULINGS					
100	120	133	150	175	200		100	120	133	150	175	200
1%	1%	1%	1%	1%	1%		94%	94%	94%	94%	94%	94%
2%	2%	2%	2%	2%	2%		95%	95%	95%	95%	95%	95%
3%	3%	3%	3%	3%	3%		96%	96%	96%	96%	96%	96%
4%	4%	4%	4%	4%	4%		97%	97%	97%	97%	97%	97%
5%	5%	5%	5%	5%	5%		98%	98%	98%	98%	98%	98%
6%	6%	6%	6%	6%	6%		99%	99%	99%	99%	99%	99%
7%	7%	7%	7%	7%	7%		100	100	100	100	100	100

Maximum three-colour printed density

Examination of the three-colour scale of the test form will reveal whether the scale is fully printable throughout its range of halftone values. The maximum printed density is likely to occur at the point in the scale which contains 100% cyan, and the printed amounts of yellow and magenta that make a near neutral step (about 88% of each). This density should be measured using the visual or non-colour filter on the densitometer, and noted. This density will be used to determine the tone reproduction and contrast range that can be supported by the printing conditions.

Maximum four-colour printed density

In the most dense area of the printed four-colour matrix, the point above which no increase in printed density takes place can be visually discerned. There is no reason for printing more ink than this point contains.

Figures 9.5a-c shows the test piece used to determine the maximum printed densities for three- and four-colour overprints (see also Plate 8).

Fig. 9.5a The four-colour overprint matrix

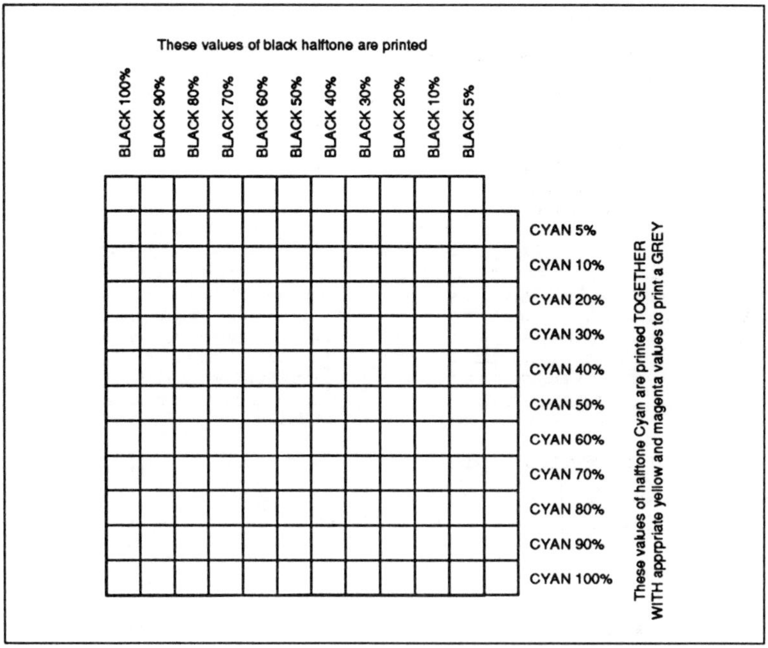

These values of black halftone are printed

BLACK 100% | BLACK 90% | BLACK 80% | BLACK 70% | BLACK 60% | BLACK 50% | BLACK 40% | BLACK 30% | BLACK 20% | BLACK 10% | BLACK 5%

CYAN 5%
CYAN 10%
CYAN 20%
CYAN 30%
CYAN 40%
CYAN 50%
CYAN 60%
CYAN 70%
CYAN 80%
CYAN 90%
CYAN 100%

These values of halftone Cyan are printed TOGETHER
WITH appropriate yellow and magenta values to print a GREY

Fig. 9.5b The black halftone values of the four-colour matrix

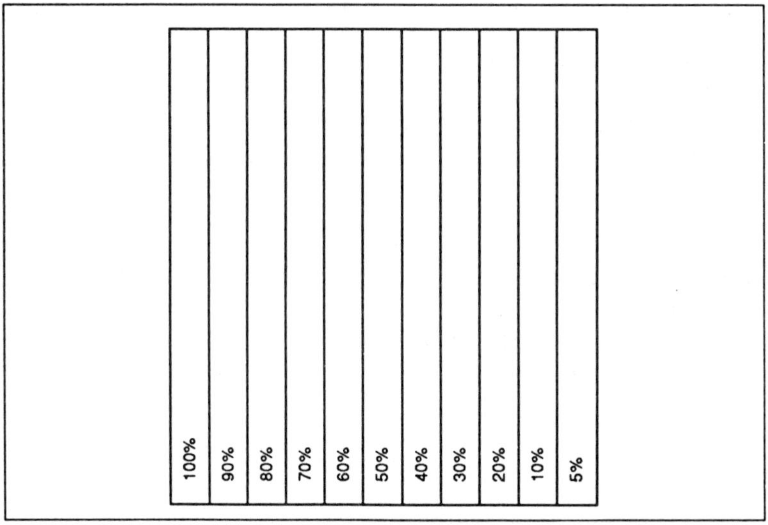

100% | 90% | 80% | 70% | 60% | 50% | 40% | 30% | 20% | 10% | 5%

195

Fig. 9.5c Three-colour target format (values given will provide an approximate grey)

CYAN 5%	YELO 3%	MAG 3%
CYAN 10%	YELO 7%	MAG 7%
CYAN 20%	YELO 14%	MAG 14%
CYAN 30%	YELO 25%	MAG 25%
CYAN 40%	YELO 35%	MAG 35%
CYAN 50%	YELO 44%	MAG 44%
CYAN 60%	YELO 54%	MAG 54%
CYAN 70%	YELO 64%	MAG 64%
CYAN 80%	YELO 74%	MAG 74%
CYAN 90%	YELO 83%	MAG 83 %
CYAN 100%	YELO 88%	MAG 88%

Colour gamut scales

The scales may be visually checked, but there are no density measurements which can be taken to convey useful data.

Application of the qualitative parameters

The application of the data extracted from the printing test form is best handled by the use of a Lloyd Jones type of quadrant diagram. This type of presentation can be used to plot a number of interdependent variables in a process, and more importantly can be used to project the characteristic of an unknown parameter.

We can say what it is we are trying to achieve (a good reproduction), and we can plot the data of the means of achieving it (the printing characteristic). We can use the quadrant to find what should be the characteristic of the remaining parameter – the halftone colour separations (see Figure 9.6).

Fig. 9.6 The quadrant diagram used to find the separation film characteristics

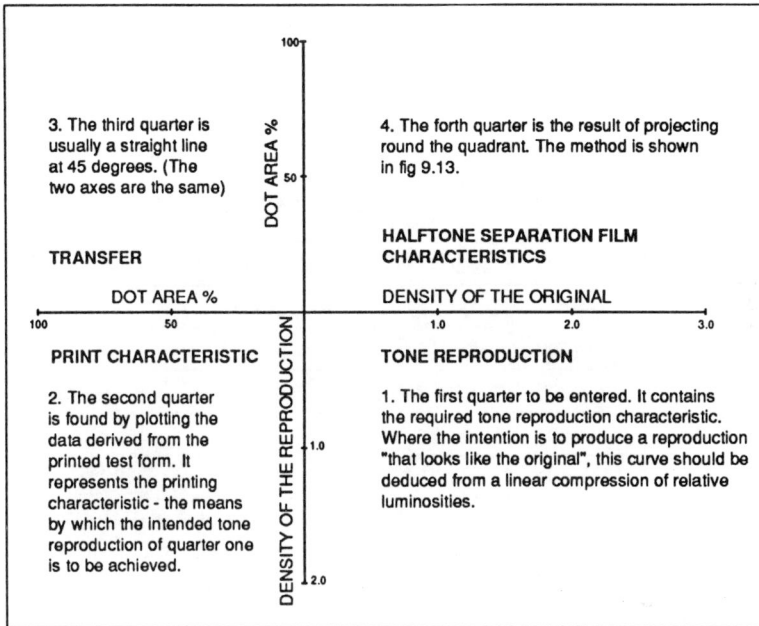

3. The third quarter is usually a straight line at 45 degrees. (The two axes are the same)

TRANSFER

DOT AREA %

DOT AREA %

100   50

PRINT CHARACTERISTIC

2. The second quarter is found by plotting the data derived from the printed test form. It represents the printing characteristic - the means by which the intended tone reproduction of quarter one is to be achieved.

4. The forth quarter is the result of projecting round the quadrant. The method is shown in fig 9.13.

HALFTONE SEPARATION FILM CHARACTERISTICS

DENSITY OF THE ORIGINAL

1.0   2.0   3.0

TONE REPRODUCTION

DENSITY OF THE REPRODUCTION

1. The first quarter to be entered. It contains the required tone reproduction characteristic. Where the intention is to produce a reproduction "that looks like the original", this curve should be deduced from a linear compression of relative luminosities.

The tone reproduction

The required tone reproduction for the whole printing process can be best be expressed in terms that are easy to measure, *ie* density. We can measure both the density of the original and the density of the printed result, and can therefore check the efficacy of the process.

Armed with the density range of the original and the density ranges of the printed test form (both three- and four-colour maximum densities are required), we can use the linear compression of relative luminosities referred to earlier, to find the compression required in the density domain. This is most easily fitted in the lower right quarter of the quadrant diagram. Two curves belong here – the three-colour tone reproduction and the four-colour tone reproduction (see Figure 9.7).

197

Fig. 9.7 Three- and four-colour tone reproduction

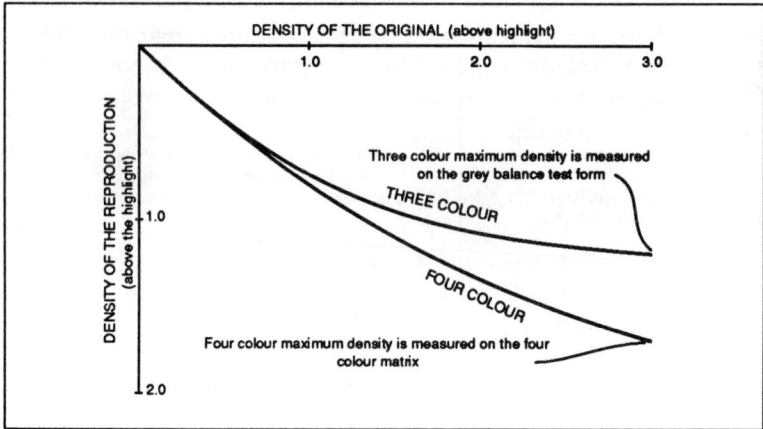

Fig. 9.7 Three- and four-colour tone reproduction

DENSITY OF THE ORIGINAL (above highlight)

Three colour maximum density is measured
on the grey balance test form

THREE COLOUR

FOUR COLOUR

Four colour maximum density is measured on the four
colour matrix

DENSITY OF THE REPRODUCTION
(above the highlight)

The printing characteristic

The halftone values from the grey sample selection are plotted, in
the lower left quarter, against the printed densities measured
from them. This renders three curves in that quarter, one from
each of the colour inks. It may be that the values for the yellow
and the magenta are the same, in which case there will only be
two curves, one for the cyan and one for the yellow and magenta
(see Figure 9.8).

Fig. 9.8 The second quadrant with print characteristic plotted for three
colours

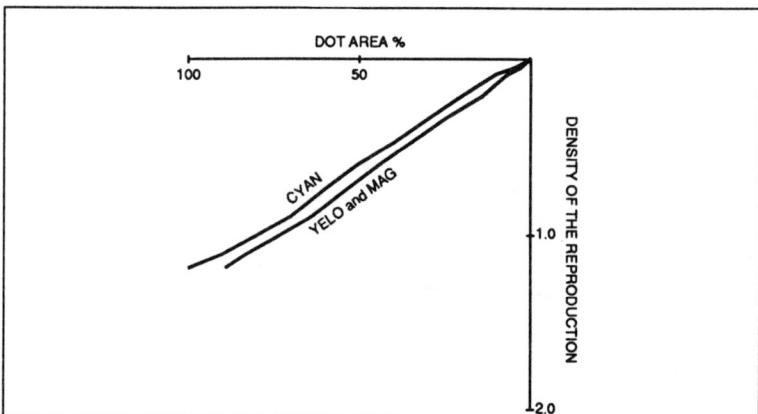

DOT AREA %

CYAN

YELO and MAG

DENSITY OF THE REPRODUCTION

198

The black print characteristic also belongs in this quarter and its derivation is somewhat complex (see Figure 9.9). It will be seen that the two tone reproduction curves in the first quarter of the Lloyd Jones diagram have the same origin and progress together for a short distance. It is where they separate that the black should begin to print.

Fig. 9.9 The derivation of the black printer

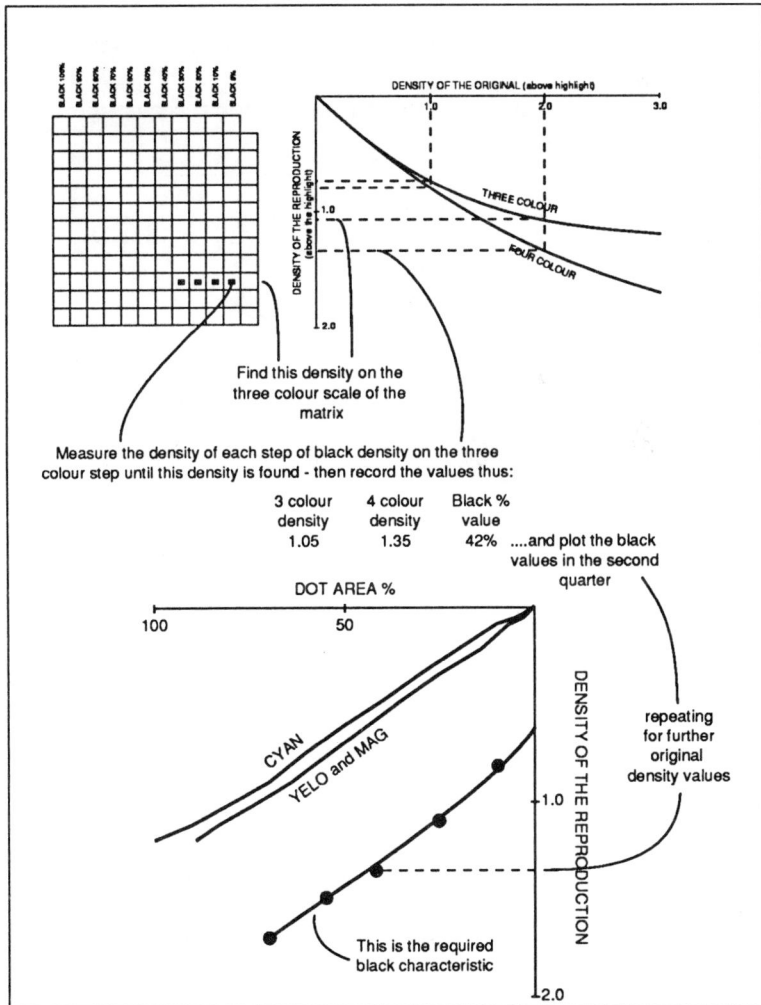

DENSITY OF THE ORIGINAL (above highlight)

1.0   2.0   3.0

DENSITY OF THE REPRODUCTION (above the highlight)

1.0

2.0

THREE COLOUR

FOUR COLOUR

Find this density on the three colour scale of the matrix

Measure the density of each step of black density on the three colour step until this density is found - then record the values thus:

3 colour density	4 colour density	Black % value
1.05	1.35	42%

....and plot the black values in the second quarter

DOT AREA %

100   50

CYAN

YELO and MAG

DENSITY OF THE REPRODUCTION

1.0

2.0

repeating for further original density values

This is the required black characteristic

199

The method for finding the tone reproduction of the black is to select an arbitrary original density and find the print density produced by three colours at that point. In the four-colour matrix, the nearest value of three-colour print density is found or interpolated between two values. At the point where the three-colour density was found in the quadrant plot, find the required four-colour density and note it. Referring back to the four-colour printed matrix, measure the four-colour density where the black overprints the established three-colour density until the new density is found or interpolated. Read off the halftone value of black that produces that four-colour density and plot it against the print density. This is the first point on a curve which will describe the black print characteristic taking the three-colour density into account.

Fig. 9.10 The four-colour print characteristic

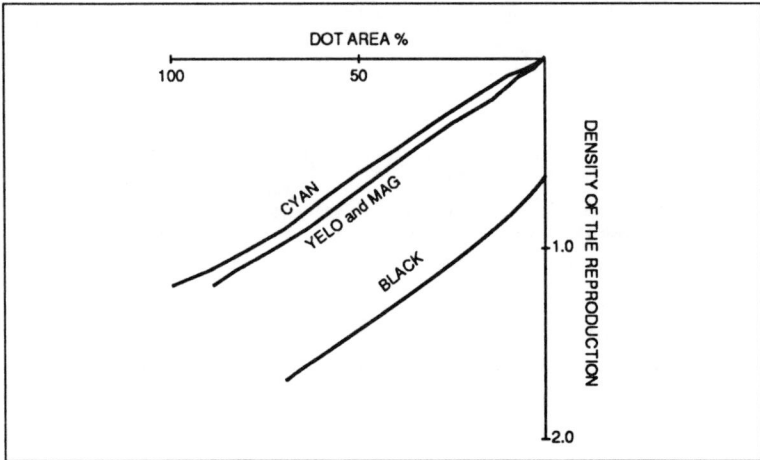

This method produces a rational black characteristic based on appearance requirements (defined by the linear compression of relative luminosities), and confirmed by the density measurements taken from the printing conditions (defined by the four-colour matrix). It therefore takes into account the dot gain of all four colours, proportionality failure, additivity failure, and the trapping errors of all four colours when they are overprinted.

The transfer quarter

The transfer quarter (Figure 9.11) is the top left on the quadrant and is a straight line at 45°, unity in other words. The halftone scales are identical.

Fig. 9.11 The transfer quarter

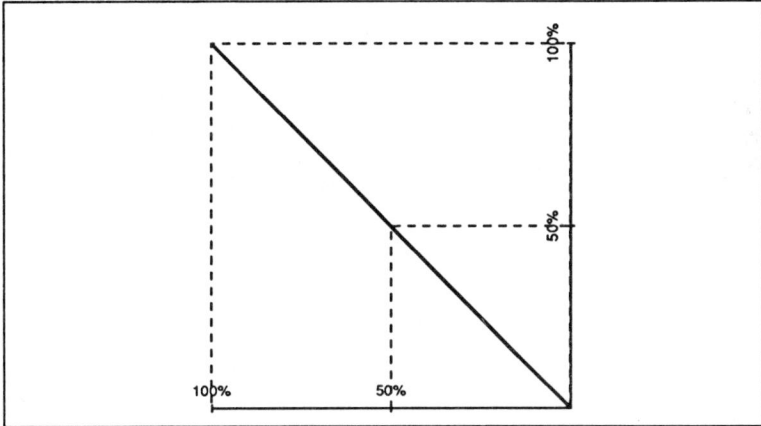

It is possible that if the test form were produced from final film and subsequent halftone separations are to be first generation halftone films, then some non-linear transfer would be implied in the transfer quadrant. Such non-linearity would represent transfer losses caused by 'soft' dots. In general, however, it is better to account for soft dots at source by suitable calibration of the conversion from density to halftone value. Some modern densitometers are programmable to allow for this conversion. Alternatively a look-up table can made quite simply.

Finding the required halftone film characteristic

The characteristic of the halftone colour separation films is a matter of projection around the three quarters of the quadrant (see Figure 9.12). By selecting an arbitrary original density, the required print density is found in the first quarter. From the print densities of each colour, the required halftone values are found. At the intersection of the halftone value and the original density

selected is a point on the curve describing the halftone separation
characteristic. The greater the number of points projected, the
greater the accuracy of the resultant curves.

**Fig. 9.12 Projecting into quarter four to find the required characteristics of
the halftone colour separations**

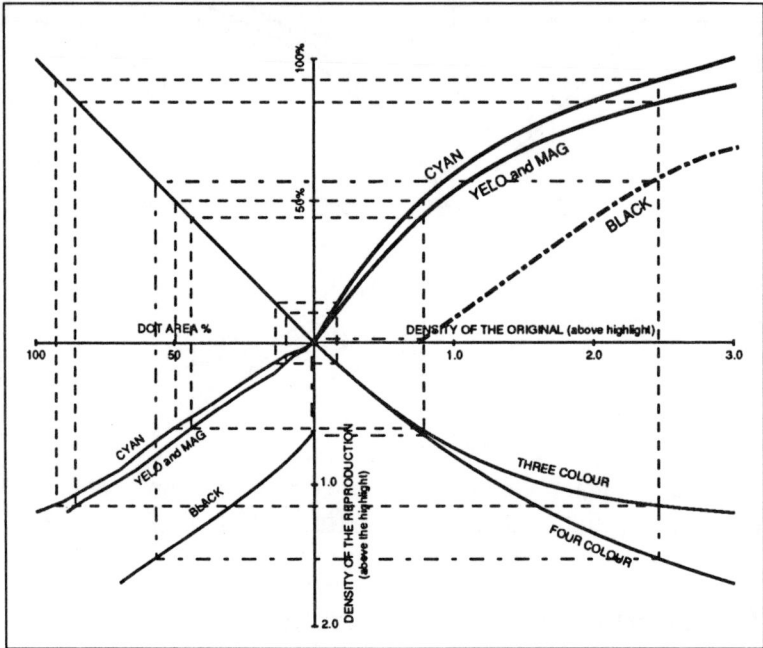

These curves then describe the characteristics that the separations
have to conform to in order to produce an accurate reproduction.
The remaining problem is to ensure that the separation system
produces such separations. We said that the vast majority of
separations are made using scanning equipment. By the simple
expedient of scanning a grey scale and making separations from
it, the separations can be measured with the transmission densitometer to check that the correct halftone values are being generated.

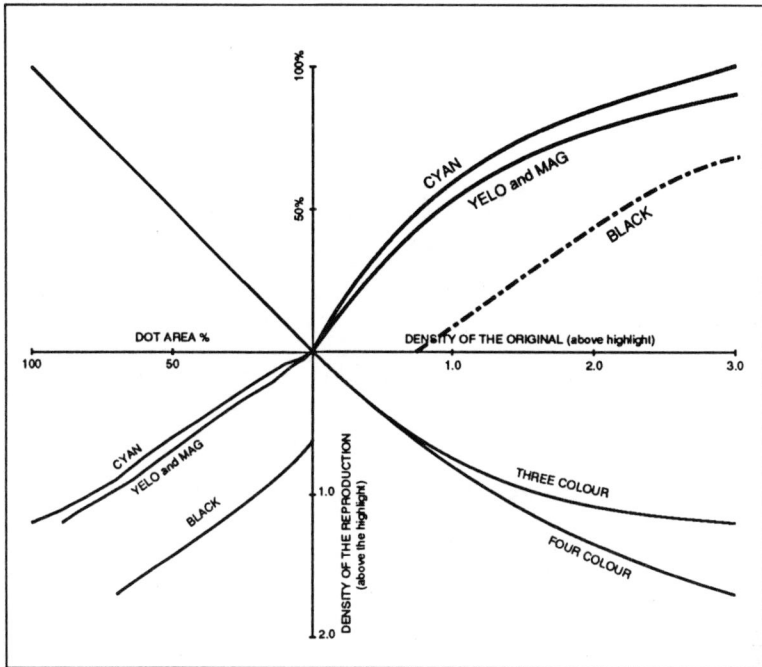

Fig. 9.13 The completed quadrant diagram showing the separation characteristics required for a reproduction which matches the original

UCR and GCR colour separations

The modern scanner is generally provided with a circuit or algorithms for the generation of these forms of colour reproduction. It is important to realise these can only work reliably if the conventional (non-GCR) reproduction is worked out on some rational basis. If it is that colour reproduction is based on an untenable method, which 'works most times', or is so contrived that the black printer, for example, covers errors in the grey balance correction, then GCR and UCR algorithms which rely on these are bound to result in errors.

Such errors may show as 'holes', or discontinuities, in the gradation of the reproduction. Grey balance correction errors will show as colour casts in the reproduction, particularly affecting greys and near neutrals.

Industry standard specifications

While this chapter has so far attempted to show how a specification for printing and proofing can be developed for an individual press, there are situations where a wider view has to be taken. This applies particularly to colour advertising in magazines and newspapers, and also some packaging work. In these cases, colour separations are usually prepared by a trade house, often for distribution to a number of printers. At the printer, separations from a number of different sources will be brought together on one set of printing plates. To avoid unacceptable compromises in these situations, it is necessary to work to some industry-wide specification.

This assumes that presses can be made to conform to the specification; the modern press can be, *provided* attention is given to getting the ink properties correct, and the press is set up properly.

There are two main specifications in use in the UK today:

FIPP – for colour advertising in magazines.

UKONS – for colour advertising in newspapers.

There are equivalent specifications in some other countries. In the US there are:

SWOP – for magazines.

SNAP – for newspapers.

The use of these specifications has advantages for everyone involved in the colour reproduction process. They have been designed to give optimum printing conditions with modern printing equipment, and have proved to be of great benefit whenever they are used.

Appendix 1 *Light sources*

Incandescent sources

These sources emit light because of their temperature. Examples are black bodies, the sun, and tungsten filament lamps.

Black bodies (or full radiators)

An example of a black body is a furnace with a very small aperture. It is black on the inside when cold, but as it is heated the inside emits visible light ranging from red at the lower temperatures, through yellow and white and blue at the higher temperatures. Thus, a particular colour can be associated with a specific temperature and this is termed the colour temperature.

The radiation emitted by these bodies depends only on their temperature and is governed by Planck's law:

$$E_\lambda = c_1/\lambda^5[\exp(c_2/\lambda T)-1] \qquad (1)$$

where E is the energy per unit wavelength, λ is wavelength in microns, T is the temperature in Kelvin (colour temperature), and $c_1 = 37415$, $c_2 = 14388$, which are constant.

We can simplify this if $T \ll c_2$ since then, $\exp(c_2/\lambda T) \gg 1$, then:

$$\lambda_{max}T = c_3 \quad \text{where } c_3 \text{ is a constant} \qquad (2)$$

This is known as Wien's displacement law, which for tungsten filament lamps is accurate to within 1%.

From formula (1) we can see that as the temperature increases E increases also. This is shown in Figure A1.1 by the increase in area under the curve as T increases. The peaks of the energy curves shift to the shorter wavelengths with increases in temperature and this shift is the displacement predicted in Wien's law.

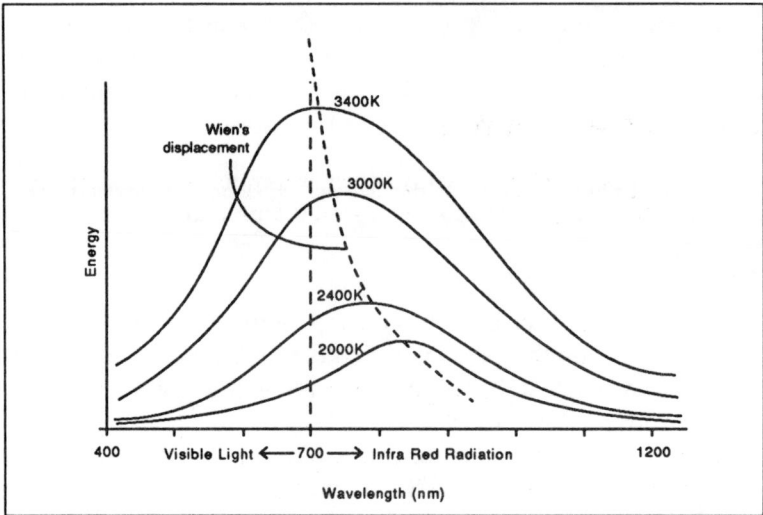

The spectral power distributions shown make no claim to accuracy but they show that the higher the colour temperature the nearer the peak becomes to the visible. In fact, at 6,500K the peak moves well into the visible and the ratio of energy in the blue end of the visible spectrum to red becomes much smaller. Also the peak becomes very much larger, thus giving a far greater energy output at any given wavelength. These curves also show that the higher the colour temperature the greater is the efficiency of the source because more visible light (as shown in Figure A1.1) and less infra-red in comparison is emitted for a given wattage. Thus tungsten, which operates at a colour temperature of 3,000K, approximately, is really quite inefficient. Also, high colour temperatures correspond to bluer light and a reduction in the difference between the colour of the light from the lamp and that of daylight. For these reasons, tungsten lamps are made to operate at the highest possible colour temperatures.

Often a more convenient method of representing the spectral distribution of the energy emitted by a Planckian radiator at various temperatures is in terms of relative energies. Thus we ignore the fact that the absolute energy of a black body at higher temperatures is greater than for low temperatures and normalise all the temperatures such that they are all equal at 600nm. We can then compare them more easily for relative energy distribution. This is shown in Figure A1.2. As we can see, at 5,000K we get a very good energy distribution. The ideal lamp would operate at a colour temperature of about 5,500K, since its energy distribution would be very flat and the energy output would be high.

Fig. A1.2 Relative energy of a Planckian radiator at various temperatures

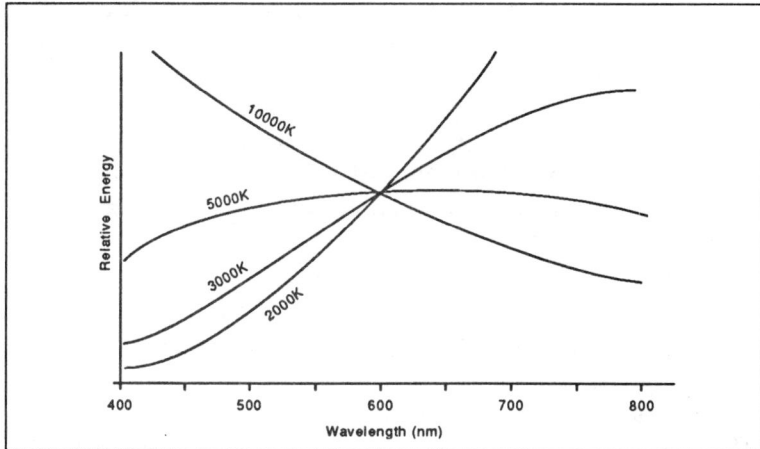

Tungsten filament lamps

These are obviously not full radiators in the sense that they do not have small openings to a heated enclosure. But the energy distributions they emit are very nearly identical to those emitted by full radiators of temperatures about 50K higher than those of the filaments, so that it has become customary to designate the colour of tungsten filament lamps by quoting these temperatures, which are referred to as colour temperatures. Thus, a lamp of temperature 3,000K, for example, emits light of spectral energy distribution almost identical with that of a full radiator operating at this temperature; the actual filament temperature would be about 2,950K but this figure is of little interest and is not generally quoted.

Since the melting point of tungsten is 3,700K it will obviously never be possible to reach the ideal temperature of 5,500K and thus one serious drawback of these lamps, apart from their overall low output previously mentioned, is their relatively low blue output. If thick filaments are used in these lamps (about one-fifth of a millimetre diameter) a temperature of up to 3,400K is possible, whereas for thin filaments (about one-fiftieth of a millimetre) colour temperatures approaching 2,700K are the maximum because of the fragility of the filament and the serious weakening resulting from any evaporation thereof. The question thus arises as to why do we use thin filaments. The problem with a thick filament is that it has a lower electrical resistance, so it can only be operated at high wattages or low voltages. Where this is attainable easily, such as in car headlamp bulbs where the power comes from a 12 volt battery direct, or in projector bulbs where a high wattage is used, then a thick filament is used. However, for domestic bulbs a thin filament is necessary to present sufficient resistance to the higher voltages used for mains electricity.

Tungsten-halogen (quartz-iodine) lamps

One serious drawback of the domestic tungsten bulb is that the tungsten evaporates and deposits on the envelope causing blackening. Also, for various reasons, this evaporation occurs more markedly at some points than others so that the filament develops local 'waists' which are thinner than the rest. Thus, the resistance is increased at these points and it gets hotter causing even faster evaporation until the filament finally breaks. This process can be retarded by running the filament in an atmosphere of low pressure iodine vapour, and building the envelope of quartz so that its wall can be maintained above about 250°C in temperature. This is maintained by making the envelope very compact. Above 250°C the evaporated tungsten reacts with iodine to form tungsten iodide which is not deposited. Above 2,000°C, however, this reaction is reversed, and so at the filament the tungsten is redeposited. Unfortunately, the tungsten does not go preferentially to the thinnest part, so the type of failure described above still occurs. However, the lamps' life is extended, or alternatively they can be run at a higher colour temperature than ordinary lamps for the same average life. Other advantages are that since they do not blacken with use, they remain more efficient and less variable during their life; they can be run at a higher colour temperature and are, therefore, more efficient; being compact they are convenient and efficient when used with optical components. One disadvantage is that iodine absorbs slightly in the yellow-green, so that if too much is used the light has a purplish tinge.

Table A1.1 gives the colour temperature of some typical tungsten lamps.

Table A1.1

Tungsten lamp colour temperatures

Lamp	K	Mireds
Domestic lamp	2,800	357
Flood light	3,000	333
Studio and projector	3,200	312
Quartz-iodine	3,300	303
Photoflood	3,400	294
St. illuminant A	2,856	350

Energy converting filters

When using filters for modifying the colour temperature of the light emitted by lamps, it is convenient to use the reciprocal of the colour temperature, rather than the colour temperature itself, multiplied by 10^6 to bring the value to a convenient size. These values are called micro-reciprocal degrees (mireds). Thus, a colour temperature of 2,000K is equivalent to 500 mireds; 4,000K to 250 mireds. A filter which raised the colour temperature of the light emitted by a source from 2,000K to 4,000K would thus produce a change of -250 mireds. Now it happens that such a filter always produces a change of -250 mireds no matter what the original colour temperature of the lamp. This is true for filters of all mired shift values, whether positive or negative, to the same accuracy as that to which Wien's radiation law is true. A filter which possesses this property is known as an energy converting filter.

Daylight

The sun, from which all phases of daylight are derived, has a surface temperature of about 6,500K. However, this light has to pass through the atmospheres of both sun and earth.

Figure A1.3 shows a typical spectral energy distribution of sunlight as received on the earth's surface. The undulations in the curve are caused by absorption bands in the solar spectrum (Fraunhofer lines) and others by absorptions in the terrestrial atmosphere (oxygen and water vapour, for instance).

Fig. A1.3 Typical spectrum of sunlight

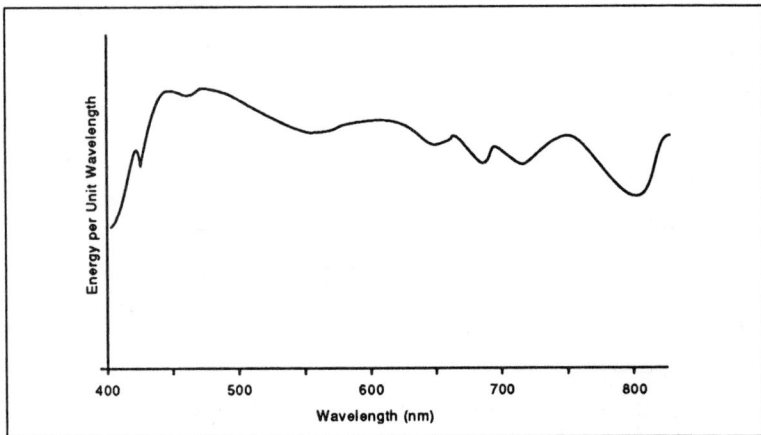

If the atmosphere is clear and cloudless, the total daylight consists of a mixture of the direct light from the sun together with the diffuse light scattered by the atmosphere. Because light of short wavelengths is scattered much more than that of long wavelengths, this diffuse sky light consists mainly of blue light and gives rise to the blueness of the sky. The diffuse light, however, is not only scattered downwards to the earth, but also outwards into space, so that there is a net loss of blue light in the combined sunlight and sky light incident on the earth, thus the colour temperature is lowered. The sun's surface probably approximates closely in energy distribution to a full radiator; but as seen from the earth's surface, because of the loss of blue light by scattering, and because of the absorptions in the atmospheres of both the

210

earth and the sun, the departures from full radiation are considerable. Figure A1.4 shows spectral energy distributions typical of
daylight when the sun is shining and when the conditions are
cloudy. For comparison with the results for a sunny condition the
curve for a full radiator of 5,630K is shown and it is seen that
while the general distribution is similar there are some appreci-
able differences. The difference between the colour temperature
to which the sunlight now approximates (5,630K) and that of the
surface of the sun (6,000-7,000K) is a measure of the loss of light
of the short wavelengths by scattering into space. Similarly, for
comparison with the results for cloudy conditions the spectral
energy distribution of a full radiator at 7,730K is shown.

Fig. A1.4 Comparison of natural light and full radiators

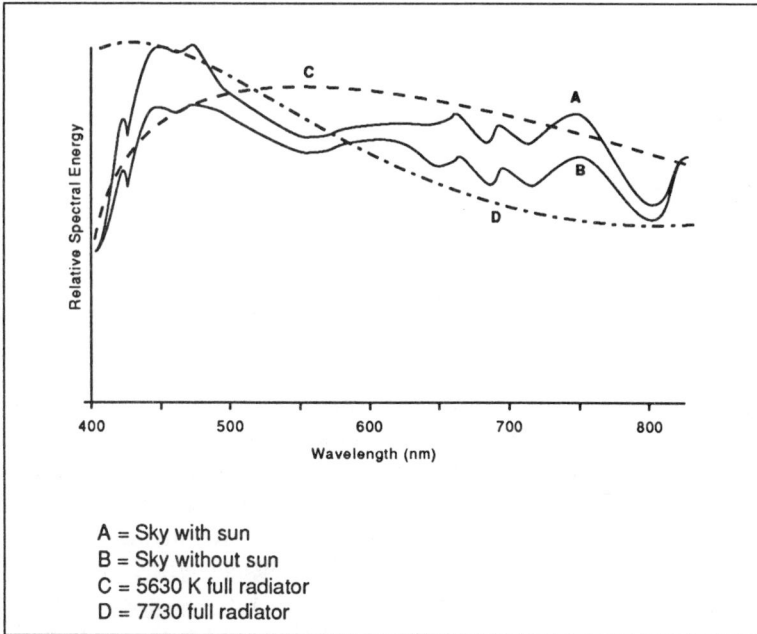

A = Sky with sun
B = Sky without sun
C = 5630 K full radiator
D = 7730 full radiator

As well as the various types of sky we get considerable variations
in spectral energy distribution depending upon the altitude of the
sun, height of the clouds, even from the colour of the ground.
Because of these considerable variations the CIE has standardised
a series of energy distributions representing daylight at all wave-

211

lengths between 300- and 800nm. Three of these, D6500, D5500 and D7500, represent standard daylight for general use, yellower daylight such as may be provided by sunlight with sky light, and a bluer daylight such as may be provided by a north sky respectively. These standard illuminants are only intended to be representative of ranges of illuminants, so that any actual sample of sunlight, for instance, might well be redder or bluer than D5500, according to the solar altitude, weather conditions, *etc.* For practical indoor viewing we have standard illuminants S_B and S_C. These are the tungsten lamp standard S_A plus filters, to represent sunlight and light from an overcast sky respectively. Both are seriously deficient at wavelengths below 400nm, however, and create a problem with fluorescent samples. Source S_C has now been largely replaced in practice by D6500.

Fluorescent lamps

If we have a mercury vapour lamp, with the gas at a low pressure, the spectral distribution consists of a series of discrete spectral lines. If the pressure of the gas is raised the energy distribution is continuous, but there are still very pronounced peaks at the same wavelengths as the discrete spectral lines with the low pressure tube. If, however, phosphors are added to this lamp, the ultra-violet light, which is quite considerable, is converted into visible and fills in, to a certain extent, the 'gaps' between the peaks. These then have a higher efficiency than the vapour lamp alone. Figure A1.5 shows the spectral energy distribution curve of a fluorescent lamp. The rectangular protuberances are caused by the lines of the mercury spectrum; the areas of these rectangles give a true representation of the amount of energy emitted at their mid-wavelengths, but their true shape is much higher and thinner than that which can be drawn conveniently on a diagram. These sharp peaks, together with the relatively small amount of light in the far red part of the spectrum can result in a certain amount of distortion in the appearance of colours.

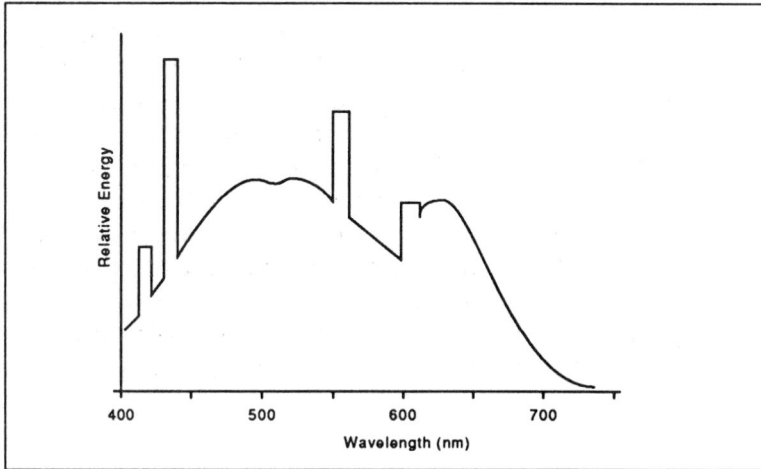

Xenon arcs

Another source providing a mixture of continuous spectrum and emission at discrete lines is the xenon arc. The exact energy distribution depends somewhat on the pressure of the xenon gas in the lamp, but it is usually fairly similar to that of daylight having a correlated colour temperature of about 6,000K. However, the emission at the red and blue ends is usually rather higher so that the light is very slightly purplish compared to daylight.

Carbon arcs

These operate in air without any glass envelope and the light produced comes from both the intensely hot craters of the carbon rods forming the arc, and partly from the combustion of gases between the arcs. The efficiency and colour of the emission are improved by incorporating additives, such as cerium, in the carbon rods. This gives a white-flame arc producing light of approximately average daylight quality. It is also possible to incorporate additives which will give arcs emitting light having a colour temperature of about 3,200K, quite close to the tungsten sources.

Correlated colour temperatures

Although colour temperature only strictly defines the relative spectral energy distribution of full radiators, it is common with other sources of 'white' light to quote their correlated colour temperature: this is defined as that colour temperature of the full radiator which produces light most closely matching that particular source. Thus, it gives an indication of the relative bluishness or yellowness of the source. Table A1.2 lists some common lamps with their colour temperatures. The corresponding mired values are also given, and this scale is particularly useful because, over the range of values involved, it so happens that equal mired intervals are to a good approximation equivalent to equal colour differences.

Table A1.2

Colour temperatures of common light sources

Source	K	Mireds
Typical north sky light	7,500	133
Typical average daylight	6,500	154
Artificial daylight fluorescent lamps	6,500	154
Xenon (electronic flash or continuous)	6,000	167
Typical sunlight plus sky light	5,500	182
Blue flash bulbs	5,500	182
Carbon arc (for projectors)	5,000	200
Cool white fluorescent lamps	4,300	233
Clear flash bulbs	3,800	263
White fluorescent lamps	3,500	286
Photoflood tungsten lamps	3,400	294
Tungsten-halogen lamps	3,300	303
Projection tungsten lamps	3,200	312
Studio tungsten lamps	3,200	312
Warm white fluorescent lamps	3,000	333
Flood lighting tungsten lamps	3,000	333
Domestic tungsten lamps (100-200W)	2,900	345
Domestic tungsten lamps (40-60W)	2,800	357

Figure A1.6 shows the most useful part of the black-body locus plotted on the u, v chromaticity diagram. For sources which do not lie on this locus the correlated colour temperature is calculated as that colour temperature whose chromaticity lies closest to the chromaticity of the source in question. Since the u, v triangle represents equal colour differences by approximately equal distances, this method of calculation gives results reasonably close to those which would be obtained by direct visual comparison by a normal observer.

215

Fig. A1.6 Various light sources plotted on the u,v colour diagram

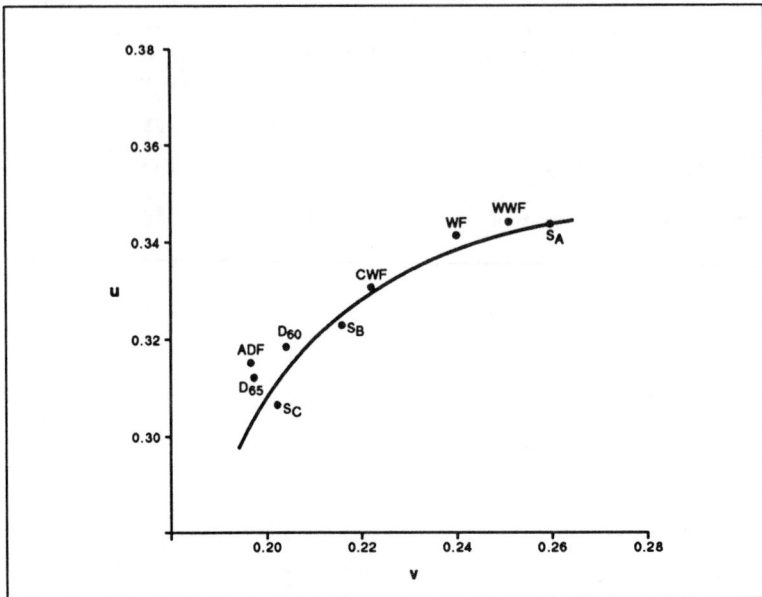

In the diagram some of the important standards are shown, plus four fluorescent tubes indicated thus: WWF – warm white; WF – white; CWF – cold white; ADF – artificial daylight.

Colour rendering

Whilst correlated colour temperature defines the colour of a source in a reasonably unambiguous way it tells us nothing about the spectral power distribution of that source. It is quite feasible to have two sources with similar colour temperatures and yet quite dissimilar spectral power distributions and the appearance of specific colours could change quite dramatically between the two lamps despite the similarity in their own colour.

Whilst specification of the spectral power distribution would be a possible method of avoiding problems arising from this, it is unnecessarily restrictive and quite impractical for lamp manufacture. Thus, two other techniques for defining the colour rendering properties of light sources have been established:

Crawford 6-band method. This is the system used for defining the sources to be used in British Standard 950 (viewing conditions for the assessment of colour) and is really an abridged form of the complete spectral power distribution definition. Basically, after measuring the spectral power distribution, the relative luminous flux at each wavelength is calculated. The visible spectrum is then divided into six bands and the relative flux present in each band specified as a percentage. Two bands are added in BS 950 to cover the ultra-violet, which is significant where fluorescence occurs.

Whilst this system has much to commend it, it can be argued that it still tells us little about the effects on any particular part of the colour space. It was to attempt to overcome this difficulty that an alternative method was recommended by the CIE.

CIE colour rendering method. This system is based on measuring the colour differences between a specific coloured sample when viewed under a reference source and a test source. Altogether 14 Munsell colours are included and the CIE co-ordinates calculated for each sample using the sources in question. Generally the reference source will be a standard spectral power distribution such as D_{65}, S_C, *etc*, and the colour measurements are calculated after measuring the spectral power distribution of the test lamp. The spectral reflectance of the Munsell samples are standard, so after measurement of the test lamp all the information required is available for conventional calculation of chromaticity co-ordinates.

Basically, for each sample the Special Colour Rendering index is computed by calculating the shift expressed by three orthogonal differences using Pythagorean means. The 1964 colour difference formula is typical. Thus:

$$DE_i = ((U^*_o - U^*_1)^2 + (V^*_o - V^*_1)^2 + (W^*_o - W^*_1)^2)^{1/2}$$

where the subscript 'o' refers to the U^*, V^*, W^* co-ordinates under the reference lamp and '1' refers to the co-ordinates under the test lamp. The subscript 'i' specifies the sample number.

The Special Colour rendering is then computed from the formula:

$$R_i = 100 - 4.6\,DE_i$$

For many purposes a General Index is also quoted and this is obtained by taking the mean of the R_i values for samples 1-8. Many lamp specifications, in fact, only include these samples, although they include no highly saturated colours or particularly important colours such as flesh, grass green or sky blue. These are included in samples 9-14. The ISO standard 3664 (the international equivalent of BS 950 pt II) uses the colour rendering index method for specification and states that a viewing surface may be considered within specification if it has a General Index greater than 90 and each special index is greater than 85, with reference D5000. However, this covers samples 1-8 only.

Appendix 2 *Introduction to logarithms*

We talk in mathematics of a number being raised to a power and by this we mean that the power states that the number is multiplied by itself that many times. For example:

$$10^2 = 10 \times 10 = 100$$

and

$$10^5 = 10 \times 10 \times 10 \times 10 \times 10 = 100,000$$

In these two examples the powers are 2 and 5 respectively. Any number may be raised to a power, for example:

$$2^3 = 2 \times 2 \times 2 = 8$$

and

$$3^4 = 3 \times 3 \times 3 \times 3 = 81$$

One of the important base numbers in mathematics is 10, and Table A2.1 lists some powers of 10 and the numbers they represent.

Table A2.1

Numbers expressed as powers of 10

Number	10	100	1000	10,000	100,000	1,000,000
Power of 10	10^1	10^2	10^3	10^4	10^5	10^6

Since $10^1 = 10$ and $10^2 = 100$, then all numbers between 10 and 100 may be represented by the number 10 raised to powers between 1 and 2. For example $15 = 10^{1.18}$ and $30 = 10^{1.48}$. If we also introduce negative numbers and zero as permitted powers then any positive number may be represented by 10 raised to a power. Some examples are given in Table A2.2.

Table A2.2

Numbers expressed as powers of 10								
Number	0.01	0.1	0.25	0.45	1	15	30	100
Power of 10	10^{-2}	10^{-1}	$10^{-0.60}$	$10^{-0.34}$	10^0	$10^{1.18}$	$10^{1.48}$	$10^{2.0}$

There are a number of advantages in considering numbers as powers of 10, one of them being that numbers can be multiplied by *adding* their powers. This can be rigorously proved but for now consider:

100 x 1,000 = 100,000

This can be expressed as:

$10^2 \times 10^3 = 10^5$

and the additive nature of the powers can be seen. Now consider the following example:

0.01 x 15 x 30 x 0.1 = 0.45

This calculation is much simpler when converted as follows into powers of base 10. These are obtained from Table A2.2.

$10^{-2} \times 10^{1.18} \times 10^{1.48} \times 10^{-1}$

The answer is obtained by adding the powers of these to obtain the answer $10^{-0.34}$ which equals 0.45 as can be seen from Table A2.2.

So, what do we mean by logarithms? Common logarithms are simply the power to which 10 must be raised in order to define any other number. Thus, from Table A2.2 the logarithm of 10 is 1 and the logarithm of 30 is 1.48. In a similar way we can define the logarithms of any number and these may readily be looked up in tables. This makes it very simple to undertake complex multiplication problems. Returning to the problem quoted earlier we showed that:

0.01 x 15 x 30 x 0.1 = 0.45

could be calculated as

$10^{-2} \times 10^{1.18} \times 10^{1.48} \times 10^{-1} = 10^{-0.34}$

In terms of logarithms we would look up 0.01, 15, 30 and 0.1 in the logarithm tables and obtain -2, 1.18, 1.48 and -1. These would be added to obtain -0.34 and this would then be looked up in antilogarithm tables to give the answer 0.45. Many calculators also include log and antilog functions.

As well as simplifying calculation, however, logarithms are very useful in defining mathematically certain physical phenomena. The way light transmission varies as a function of the amount of silver or ink pigment present is an example of this and it explains why density is so appropriate.

Index

A

absorption 8, 70, 81, 90
    unwanted 73
achromatic 99
actinic light 12
adaptation 25
additive colour mixture 28, Plate 1
    see also colour mixture 81
additivity 167
    additivity failure 176
Amberlith 121
analogue scanner 107
argon-ion laser 113
Automask 121

B

back trapping 160
balance 84
balanced hue 85
banding 112
beam computers 113
black 30, 70, 98
black printer characteristic 199
blanket 2, 141
blue key 123
blue shift 58
brown 84
Brunner 134
BS 950 part 1 55
BS 950 part 2 55
burn-out 11, 121

C

camera format 56
camera ready copy 51
catchlights 49, 64

dye transfer print 48, 53
dynamic range 109

E

EDG 36, 116
electro-magnetic waves 21
electronic domain 101
electronic dot generation 36, 116
electronic flash 62
electrophotography 150
emulsion stripping 51
end density 50
exact reproduction 93
expose section 103
exposure 56, 64
extra colours 77
eye 23, 34, 57

F

facsimile reproduction 53
film
    assembly 10, 120
    size 56, 58
    stock 56
filmless repro 101
filters 77, 81
    ideal 78
final film 122
flash 61
flat-bed 137
flat field scanner 108, 154
flats 11, 122, 131
flesh tones 57, 64
flexography 1, 2, 120
fluorescent light 53, 62, 212
FOGRA 134, 164
foil 11, 122, 129
fringes 111

G

gamma 101

imposition 128
incident light 153
infra-red light 57, 58, 206
ink
    colour 181
    deficiencies 86
    density 77, 141, 156
    film thickness 81, 85, 91, 157, 160, 182, 185
    ideal 71, 79
    jet 5, 148
    real 81, 82, 86
    rheology 143
    sequence 143, 181
    set 67, 82
    temperature 159
intaglio 1, 3
intensity 61
internal drum 115
ISO 3664 49, 55, 192

K

key 123

L

laser 113, 131
lateral geniculate nucleus 25
layout 10, 11, 121, 122
letterpress 1, 2, 120
light
    nature of light 20
lighting 56, 205
    choice of lighting 61
    mixed lighting 61, 63
lightness 74
linear array 108
litho blanket 2, 138
litho platemaking 131
lithography
    see offset lithography
Lloyd Jones diagram 93, 196
local contrast 110

logarithm 45, app. 219
look-up tables 87
low key 49
luminosity 48, 93-98

M

make-ready 3, 123, 125, 165
mask 121
masking 83, 88
matching stimuli 28
mechanical dot gain 182
metameric 26, 82, 85
metamerism 27, 54, 80, 82
   object metamerism 26
mid-tone 64
mired 61
mixed originals 54
modulator 113
moiré 119
monochrome 33
montage 51, 120
mounting foils 123
Murray-Davis 158, 167

N

negative stimulus 29
nerve fibres 24
neutral greys 57, 84, 91
Nielson 158, 176
noise 109
non-linear masking 90
non-reproducible colours 76

O

offset lithography 2
oil painting 7, 48, 54
op-amp 104
opacity 45, 182
operational amplifier 104
opponent colour theory 25
optic nerve 24

optical brightener 58
optical density 171
optical dot gain 158, 182
optical geometry 155
optics head 113
optimum ink level 185
original 48
overlay 53

P

page ready copy 51
pass sheet 160
peelable membrane 121
phosphors 28, 70
photo-diode 108, 156
photo-mechanical process 34
photo-opaque 11
photo-sensors 108, 156
photo-sites 109
photocell 152, 155
Photofloods 61
photographer 56
photographic colour prints 48, 52
photographic domain 104
photographic masking 8
photographic montage 51
photomultiplier 103
photopolymer 138
picking 144
picture detail 118
picture elements 118
pigments 139
pin register 123
Pira 67, 163
planning 10, 11, 123, 128
plate transfer characteristic 183
plate/foil punch 123
platemaking 131, 144
polarised light 174
polyester 122, 129
polymer 131

subtractive colour mixture 30
   see also colour mixture 81
surface finish 52
surround 25, 55
synoptic layers 24

T

target 168
target proof 139
television 28
test form 141
texture 22
thermal sublimation 149
thermal wax transfer 149
three dimensional colour space 74
tint blocks 11
tolerances 163, 181
tone
   compression 48
   range 53
   reproduction 86, 91, 141, 169, 192, 197
   value 156
TRAND 92
translucency 22
transmission 45, 151
transparency 48, 53, 58, 64
   ideal 49
trapping 141, 156, 167, 176, 183
tungsten 61
   overrun tungsten 61
   tungsten halogen 61
typesetter 114
typographical proof 138

U

UCR 87, 99, 183
ultra-violet light 12, 57, 132
undercolour removal 87, 99, 183
uniform visual scale 93
unitary hues 23